Brian Bagnall

VARIANT PRESS

Designed by Relish Design Studio Ltd.
Original artwork by Wil Glass
Editied by Sylvia Philipps
Cover by Karen Armstrong
Printed in Canada

Library and Archives Canada Cataloguing in Publication

Bagnall, Brian, 1972-
        Maximum Lego NXT : building robots with Java brains / Brian Bagnall.

Includes index.

ISBN 978-0-9738649-1-5

        1. Robotics--Popular works. 2. LEGO toys. 3. Java (Computer program
language) 4. Robots--Programming. 5. Robots--Design and construction. I. Title.

TJ211.15.B34 2007   629.8'92

C2007-902135-2

VARIANT PRESS
143 Goldthorpe Crescent
Winnipeg, Manitoba
R2N 3E6

Foreword

When we launched LEGO® MINDSTORMS™ back in August of 1998, we did not anticipate the immense attention it would receive. What was internally (for a time) referred to as the *shadow market*, the adult fans of LEGO, was not our intended target group.

Little did we know the impact these fans would have on our technological flagship – our first full venture into connecting the world of LEGO bricks with the booming world of computers.

Within days after the Robotics Invention System 1.0 hit the shelves the RCX was laid bare. Detailed pictures of its innards were posted on the Internet for all to see, complete with a shopping list of its components. The first attempts at hacking the software had begun.

Our legal department was soon on alert, not knowing how to react to what they felt was an attack on our intellectual property! After many discussions, we decided not to impose sanctions on those who cracked and hacked our product; it appeared there were no actual legal transgressions. We bided our time and watched how these fore-runners actively used the Internet to share knowledge. Eventually they developed alternative programming languages and applications for the RCX. We saw how this actually created a strong and vibrant buzz around LEGO MINDSTORMS.

These pioneers have taken LEGO MINDSTORMS to levels that leave those of us developing the product at LEGO in awe. Thousands of models have been featured on the Internet and at fan-driven events. A multitude of home-brew sensors and similar applications have been made and publicized. Whole communities have sprung up, along with fans and proponents of various programming languages.

It is the pioneering spirit and enthusiastic involvement of such communities that has made LEGO MINDSTORMS into what it is today, and for that we are grateful. We want to nurture this strong interest in our product and maintain the dynamic relationship we have established between our company and the users.

Our awareness of this strong core of adult users with their deep understanding of software, hardware and firmware, led us to consider involving these users in our innovation process. We realized that there was enormous potential from these experts who knew so much; maybe more than we did! Why not make the most of it?

We knew that children in the 10 to 14 age group would be limited in suggesting what they might want in a new generation of LEGO MINDSTORMS. They would have only the old generation to refer to. However, the programmers and amateur sensor developers knew what it would take to implement new ideas.

We have to admit that we did not ask the top management for permission to actually invite users inside our R&D facility, fearing they would say no, but when we revealed what we had done there was praise all round for the initiative! And it has set a precedent for how we want to involve users in our innovation process.

By January 2006, we felt we had created a strong product. At the Las Vegas Consumer Electronics Show (CES) we announced the advent of the LEGO MINDSTORMS NXT robotics tool set, to be launched in August of the same year.

The response to our announcement was exactly as we hoped. Many of the people behind the alternative languages for the RCX picked up the torch to port their applications to the NXT brick.

We constantly strive to raise awareness about LEGO MINDSTORMS, and books have been an important part of this. More than 40 books have been written about the first generation RCX. As of this writing (spring 2007) there are already books about LEGO MINDSTORMS NXT on the streets. This book is one of the best examples of them, and I hope you enjoy building and programming the LEGO robots within.

—*Søren Lund, Director of LEGO MINDSTORMS*

Contents

Acknowledgments

Thanks to Lawrie Griffiths, Professor Roger Glassey, and Charles Manning for their incredible work on leJOS NXJ. Thanks to Mark Klein, David J. Perdue, and Philippe "Philo" Hurbain for their assistance with the rendered instructions. Thanks to Steve Hassenplug and Bryan Bonahoom for allowing me to use their ingenious differential in this book. And a big thanks to LEGO for coming through with another great product!

—*Brian Bagnall*

Preface

This book will explore how to build robots that recreate human abilities. We will use LEGO® NXT to build hands, arms, and legs. We will also make robots that see, hear and feel their environment.

The human body is remarkable, but there are things machines can do that humans cannot. For instance, machines can roll around on wheels and communicate through radio waves. We will take advantage of these super-human features to create even more advanced machines.

Although this book contains many projects, it is not a project book; it is a concept book. Each chapter presents a key concept that will allow you to make your own incredible robots.

If you aren't familiar with LEGO or Java programming, don't worry. This book does not assume you are experienced in any of these technologies. As you go through the chapters, you will learn new concepts and build on what was covered in previous chapters.

In some cases, we need complex software to explore subjects like voice recognition or image analysis. In these cases, I've tried to find the best open source software available.

The robots in this book use 100% NXT parts. You will be able to build every robot in this book with the LEGO parts included in the NXT kit. There is also a wealth of excellent third party sensors available for NXT robots, and it would be a shame not to explore the added functionality they bring to robotics. I've included these sensors where appropriate, often as an upgrade to the basic NXT project.

For those who want to try something adventurous with their robots, there are projects that use non-LEGO devices. One chapter uses a wireless video camera to monitor the environment. Another chapter uses a data glove to control a robot. I believe these are among the most exciting projects in the book.

A Chinese philosopher once said:

Tell me, I'll forget.
Show me, I'll remember.
Involve me, I'll understand.

That is the philosophy of LEGO and also of this book. Thanks to the ingenuity of LEGO NXT, you won't have any problem becoming involved. Let's get started.

Meet NXT

Topics in this Chapter

- A Brief History of MINDSTORMS
- Discovering the NXT Kit
- Kits and Accessories
- Surveying the Competition

Chapter 1

MINDSTORMS is both a toy and an engineering tool. It may sound like an exaggeration to place it in the field of engineering, but there are many examples of professional engineers using it to rapidly prototype inventions. I've seen it used to design everything from movie special effects to a barbecue mechanism.

As a toy, there is nothing like it. Most toys do one thing and soon lose their novelty. MINDSTORMS is always new. Owners can create a unique invention, play with it for as long as it holds their attention, then destroy it and reshape it into something new. It is a toy that never grows old.

LEGO released the MINDSTORMS Robotics Invention System (RIS) in 1998, causing a tidal wave of interest in the robotics community. Since then, owners have anxiously awaited the next step in the evolution of MINDSTORMS. Was it worth the wait? This chapter will attempt to answer that question.

A Brief History of MINDSTORMS

In 1987, the MIT Media Laboratory began developing a device they called the Programmable Brick, under sponsorship from the LEGO Group. The main designer was Fred G. Martin, and between 1987 and 1998 he turned out several different versions of the Programmable Brick. The final version is a big red brick with four output ports for motors and six input ports for sensors (see Figure 1-1).

Figure 1-1 The MIT Programmable Brick

Figure 1-2 The RCX brick

The Robotics Invention System soon followed. The heart of the RIS Kit was a yellow RCX brick that some felt bore a resemblance to Spongebob Squarepants (see Figure 1-2). It was modest compared to today's technology, using an 8-bit processor at 16 MHz and a scant 32 kilobytes of memory – less memory than the Commodore 64 had in 1982. Even with these limitations, MINDSTORMS users built incredible machines.

More than a million people said, "This is what I have been waiting for" and purchased an RIS kit. The MINDSTORMS community is now the largest robotics community on earth, easily able to share their robot designs and code with others.

Many schools embraced the technology as a learning tool. Now that the hardware has been further refined, it is even more suitable for education. There is no underestimating the effect this technology will have on kids. As more schools adopt MINDSTORMS, we may see a robotics boom in the coming years as these students enter the business world.

University students have also embraced LEGO, using it for research and graduate projects. Hopefully NXT will spur students to create even bolder designs.

The big surprise for LEGO was the hacker community. Hackers soon unlocked the RCX brick and began using alternate programming languages such as Java. Many attribute the success of the RIS kit to the availability of these programming languages.

LEGO has tried to expand the user base of MINDSTORMS. Roughly 70% of MINDSTORMS owners are adults, with an average age in their mid-thirties, so it seemed natural for LEGO to want to expand ownership to their traditional younger audience. Their strategy was to simplify the technology (by removing features) and sell it for less. The result was the Scout and later the Spybot (see Figure 1-3).

In 1998 LEGO released Cybermaster, a kit similar to the RIS kit. This contained true radio communications with a PC, similar to the NXT, except that the brick was less advanced than the RCX (see Figure 1-4).

Figure 1-3 Scout, Micro-Scout and Spybot

Figure 1-4 The Cybermaster
brick and transmitter

Between 1998 and 2006 there was a long wait for a sequel. Unknown to MINDSTORMS users, LEGO began work on a sequel in early 2004. The project was led by a brilliant Dane, Søren Lund who eventually created a robot brick that used standards such as Bluetooth and I²C (more on these later). When development began, Lund hung a sign in the development lab: "We will do for robotics what iPod did for music".

On August 2, 2006 LEGO rolled out the NXT kit. Will the kit exceed the popularity of the previous generation? If the technology inside the NXT brick is any indication, it will be.

Discovering the MINDSTORMS NXT Kit

The NXT kit has a look that says the future is right here in this box. Every part is molded from the same strong plastic used for LEGO blocks, so that when a robot starts to rebel it won't destroy itself. The sensors are molded in grey and white with orange highlights. This styling makes the robot parts look like something out of the cartoon *Robotech*™. The kit takes the seemingly difficult art of robot building and reduces it to something a child can manage. Let's examine the major parts inside the kit.

NXT Intelligent Brick

The new brick, which LEGO calls the *NXT Intelligent Brick*, looks much more refined than the yellow RCX brick. In fact, the style looks inspired by the iPod in many ways, such as color and menu navigation (Figure 1-5). Think of it as an iPod for motors.

The NXT brick is very durable. I witnessed a heavy LEGO robot fall square in the middle of the LCD screen without damage (although I don't recommend trying this yourself). Several times I drove my robot off a table, and though LEGO parts flew off, nothing broke.

Figure 1-5 The NXT Brick next to an Apple® iPod™

Although technology generally becomes smaller with each generation, the NXT is actually slightly larger than the RCX brick. The RCX brick was 6.5 cm by 9.5 cm, while the NXT is 7.2 cm by 11.2 cm. With six AA batteries, NXT weighs 286 grams (160 grams without batteries), slightly more than the 280 gram RCX.

Bigger is sometimes better, as is the case here. The NXT brick contains an Atmel® 32-bit ARM processor running at 48 MHz. This processor has direct access to 64 KB of RAM, and 256 KB of flash memory. To put this in perspective, the NXT has ten times more memory than the RCX. The flash RAM stores programs and data even when there is no power, which saves battery life.

In an age of gigabytes, you might think that an increase of mere kilobytes is not impressive. However, there are a few things that make this memory limitation irrelevant. First, robots don't use heavy-duty graphics or sound, which consumes heaps of memory in modern computers. Only mapping projects and advanced artificial intelligence (AI) programs use significant memory. For a good example, look at the specs for the Mars rovers *Spirit* and *Opportunity,* launched in 2003. Each contained 256 KB of flash memory. If that's enough memory for NASA, you can trust it is enough memory for your own projects. If you do need a lot of memory for a program, you can control the NXT wirelessly from code running on your personal computer.

The brick also contains an Atmel 8-bit AVR processor running at 8 MHz to operate the servo motors and rotation sensors. This processor has access to an additional 4 KB of FLASH memory and 512 bytes of RAM. Why a separate processor just for the motors? Because the NXT needs to monitor the optical tachometer constantly in order to remain accurate. If this task was left to the main processor, it might fail to keep track of rotations while simultaneously doing something else, such as running your program.

Batteries

The NXT brick uses six AA batteries which provide 9 volts. However, if you use your NXT a lot, disposable AA batteries will end up costing you a fortune. There are two off-the-shelf options for rechargeable batteries.

The first option is the rechargeable lithium ion battery from LEGO (see Figure 1-6). This battery provides at least 7.4 volts (closer to 8.2 volts after recharging). It fits into the regular battery case, but it also increases the depth of the NXT brick slightly.

LEGO also sells an AC adapter for charging the lithium ion battery which, conveniently, can be done while it is still inside the NXT brick. Conceivably, this means you could devise a robot that drives up to a recharging station when the batteries are low!

Lithium batteries provide power even while they are being charged, meaning your robot can also feed directly from household current. People who want to create robots that operate 24 hours a day, seven days a week (such as an Internet controlled robot) will find the lithium battery a necessary accessory (see Appendix A for pricing and availability).

Figure 1-6 LEGO Lithium Ion battery

Another option is to use six rechargeable AA batteries. There are two kinds: Ni-MH (Nickel Metal Hydride) and Ni-Cd (Nickel Cadmium). Both work well, but the Ni-MH battery supplies 1.2 volts while the Ni-Cd battery supplies 1.25 volts. They don't store as much charge between recharging as lithium, however (see Table 1-1).

Warning: Rechargeable batteries provide 7.2 to 7.5 volts to the NXT, which means your motors will not operate as fast or powerfully as they would with 9 volts. Other than that, rechargeable batteries work well.

	Rechargeable?	Volts	Duration
Alkaline	No	9.0	longest
Ni-Cd	Yes	7.5	lowest
Ni-MH	Yes	7.2	40% more than Ni-Cd
Lithium	Yes	7.4	2 x Ni-Cd

Table 1-1 Comparing battery options

Speakers

Sound is fantastic compared to the RCX. The NXT contains a sound amplifier chip that can play sampled sound. You can even make your own recordings and upload them to the NXT brick. Don't expect this to sound like MP3 quality, however. Because of memory limitations, the NXT can only play low fidelity sound with a low sample rate. LEGO has also included a library of sounds, including some robot sounds to enhance your creation.

TRY IT: *LEGO has included an on-board program to demonstrate the built in tachometer in the motor.*

1.1   Plug a motor into Port A.

1.2   From the NXT brick main menu, select View > Motor Degrees > Port A.

3.3   Turn the motor with your hand and listen to the results.
The speaker can produce sounds only a dog could hear.

Input and Output Ports

The NXT has four sensor ports and three motor ports – one more sensor port than the RCX. The sensor ports use a standard protocol called Inter-Integrated Circuit or I²C (pronounced I-squared-C). I²C is a bus that allows transmission of data to and from sensors. Philips invented the standard in the early 1980s and since then it has seen some use in cell phones and other small devices.

Although there are four physical ports, the I²C ports are capable of using far more than four sensors at once. As long as the sensor uses Auto Detecting Parallel Architecture (ADPA), you can connect additional sensors using an expander (see Appendix A).

LEGO has also included one high speed communications port in sensor port 4. This port, called an EIA-485 link, is capable of transmitting data at 926.1 Kbits per second (compared to 460.8 Kbits per second for Bluetooth). What kind of devices will use this port? Think digital cameras, sound recording devices, and possibly memory expansion.

Cables

The old RCX cables contained two wires but the NXT cables contain six wires. This makes the new cables more rigid. The NXT kit contains seven cables (one for each port) with three different lengths (see Table 1-2). If any of these prove too short, there are other lengths available online (see Appendix A).

Short	Medium	Long
20cm (8.5 inches)	35cm (14.5 inches)	50cm (18.5 inches)

Table 1-2 Comparing cable lengths

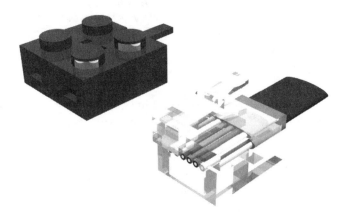

Figure 1-7 RJ12 connector next to the RCX connector

LEGO NXT uses a connector known as RJ12, which looks much like a phone connector (see Figure 1-7). Since the connectors for sensors and motors are identical, you might think you can hook up a motor to a sensor port. This does not pan out, unfortunately, because the wire signals are different.

RCX connectors plug in just like LEGO bricks. They were novel, but if you rotated the connector 90 degrees the motor rotated in the opposite direction, which was often confusing. There is no chance of improperly plugging in a cable with RJ12 connectors.

LCD Display

The NXT display is dramatically improved. The RCX contained LCD segments, which meant it could only display icons and numbers, like a calculator display. The NXT display is far more robust, allowing bitmapped images. It has a resolution of 100 x 64 pixels and an area of 26.0 mm x 40.6 mm.

The display is black and white only, with no intermediate grey. The LCD requires 17 ms to draw a new screen. It can refresh the display almost 60 times per second (60 Hz) and easily displays animations such as the introductory animation when the NXT it first powered on. Several games have already been created for the NXT, such as Tetris™ and Centipede™ clones.

The menu system on the NXT is brilliant. You can access any number of functions using the four buttons: two light grey arrow keys, an orange *enter* key, and a dark grey *back* key (see Figure 1-5). The functions contained in the menu system range from Bluetooth settings to playing sound files.

TRY IT: *With the LEGO NXT, you don't even need a computer to program a robot. Instead you can use the menu system to write a very simple program with up to five commands. For this example we will create a program for the Tribot (see LEGO kit) that reverses and turns when it hits an object.*

1.1   *Connect motors to ports B and C.*

1.2   *Connect a touch sensor to port 1.*

1.3   *From the main menu, select NXT Program (the display tells you which ports to plug motors and sensors into). Press the orange button again.*

1.4   *In box one, select the Forward icon.*

1.5   *In box two, select the Touch icon.*

1.6   *In box three, select the Back Right 2 icon.*

1.7   *Leave box four blank. In box five, select the Loop icon.*

1.8   *Now you can run the program, and you can even save it with a new name. If you decide to save the program you can access it from* My Files > NXT Files.

Bluetooth and USB

The RCX had one proprietary method to upload code using an IR tower. This was often unreliable, slow and somewhat awkward to use. The NXT provides two standard methods to transfer code: USB and Bluetooth. Both are fast and reliable.

USB

USB can be used to upload code and data to the NXT, and it is the only method for uploading firmware (more on this later). The USB port can transmit data at 12 Mbits per second. This solution is familiar to most computer users and easy to use. LEGO even supplies a standard USB cable which is identical to a printer cable.

 NOTE: *When your NXT is plugged in using the USB cable, you can view it in the Device Manager for your operating system.*

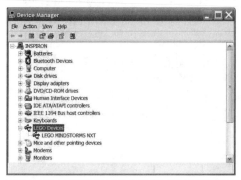

Figure 1-8 LEGO NXT in Devices

Bluetooth

Many people are not familiar with Bluetooth since it has yet to catch on in a big way. What is Bluetooth? Briefly, it is a wireless connection, like a USB port without cables. It is ideal for computer peripherals like keyboards and mice.

What isn't Bluetooth? Well, for starters it isn't the same as Wi-Fi (otherwise known as IEEE 802.11), which is used for networks. Wi-Fi is more popular than Bluetooth, but it would be overkill for a device like the NXT. It also requires more computing resources and more power to maintain a connection, which drains the batteries faster.

Bluetooth (otherwise known as IEEE 802.15.1) operates on nearly the same frequency band as Wi-Fi (2.45 GHz vs. 2.4 GHz). However, you will probably not experience any problems even if you use both on the same computer, since they automatically change frequencies. Bluetooth transmits data at 460.8 Kbits per second – 26 times slower than USB and more than 400 times slower than 802.11n.

TRY IT: *The LEGO NXT allows you to upload code directly from one brick to another using Bluetooth. You will need a second brick to try this. Go to My Files > Software Files and select a program. Click the right arrow button to scroll to the Send icon. By selecting this you can choose which NXT in the area to upload the program to.*

To use Bluetooth, you need a Bluetooth dongle (see Figure 1-9). The NXT kit does not include a dongle since Bluetooth isn't required to upload programs to the NXT. Do you really need one of these? The short answer is yes, go right now and purchase a Bluetooth dongle (see Appendix A). You can find Bluetooth dongles in electronics stores, from LEGO, and on eBay.

WARNING: *Make sure to buy a Bluetooth 2.0 dongle as the NXT does not work with Bluetooth 1.2 and older versions. Versions higher than 2.0 will also work, because Bluetooth is backward compatible.*

Figure 1-9 Bluetooth dongle

Bluetooth offers a solution if you want to create a program that uses more than 256 KB, such as video analysis, mapping, voice recognition, heavy duty AI code (chess for example), or something that uses a database. Bluetooth effectively gives you access all the memory in your computing device. This new paradigm means programmers can put the reflex actions on the NXT brick and the brains on the PC, accessible via Bluetooth.

Bluetooth also allows robots to interact with resources a computer interacts with, such as a webcam, databases, other software packages, your network, the Internet, eMail (which can interact with pagers), a web page, a printer – the list never ends. Basically the Bluetooth dongle will open up incredible possibilities for what you can do with your robot.

Bluetooth particularly shines when compared to the RIS IR tower. The RCX required a direct line of sight between the IR port and the IR tower. You could control the robot remotely but the IR had limited visibility; if the robot faced the wrong way the connection was severed. With wireless Bluetooth you can control the robot no matter what direction it faces. The only limitation is distance, which is limited to 10 meters with class II Bluetooth dongles (though most work up to 100 meters).

Chapter 7 will explore Bluetooth in more depth.

Motors

Proprioception is the ability to keep track of internal conditions of body parts. When you move your arm, you know its position even with your eyes closed. Robots can also have proprioception by sensing the position of the motor axle. The old RIS motors could not tell the position of the motor axle, so LEGO released separate rotation sensors. Unfortunately, each sensor used one sensor port, leaving no free sensor ports if a robot used three motors.

The new motor uses built-in tachometers to keep track of axle rotation. It can turn in any number of directions for thousands of rotations and come back to the exact starting position at any time. This new feature opens up incredible possibilities for robot creation, especially with navigation and robot arms. Even better, all four sensor ports remain free.

LEGO generously includes three motors in the NXT kit instead of the two offered with the RIS kit. Each motor weighs 81 grams, almost double the 42 gram RCX motors.

The NXT *servo motor* contains gears that reduce the speed of the motor and increase power (see Figure 1-10). This gives the motors an unusual shape, making them look like R2-D2 legs. Sometimes this shape can make it more difficult to integrate motors into your design, but often it can help form the structure of a robot. The old RIS motors had a short

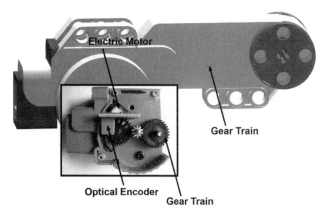

Figure 1-10 Inside the LEGO NXT Motor (photo by Philippe Hurbain)

protruding axle. The axle is now inverted into an axle hole. This makes it easier to choose the appropriate axle size.

RCX motors had only one motor speed (albeit with different power settings, which didn't consistently alter speed). Robot builders had to rely on gear reduction to attain the desired motor speed. With the NXT, motor speed can be changed through code, thus requiring fewer gears.

Sensors

The sensors in the NXT kit are identical in size and shape, except for the distance sensor which looks like a robot head (the distance sensor is wider at the front, but the rear is identical to the other sensors). All of the sensors use the same means of attaching to the robot (three holes for friction pins or axles), making the sensors interchangeable.

Touch Sensor

The touch sensor is the most basic sensor in the NXT kit (see Figure 1-11). It has a simple switch activated by the orange button on the front. The button has an axle hole, allowing you to connect parts directly to the sensor switch.

Figure 1-11 Touch sensor

Figure 1-12 Light sensor

Light Sensor

The light sensor measures the intensity of light entering a tiny lens on the front of the sensor (see Figure 1-12). The sensor is also equipped with a red light-emitting diode (LED) which illuminates the scene in front of the sensor. The sensor can also detect light invisible to the human eye, such as infrared (IR) light emitted from a television remote control.

TRY IT: *The LEGO software has a demo program for the light sensor that we can use to detect IR light from a TV remote control.*

1.1   Connect the light sensor to port 3.

1.2   From the main menu on the NXT brick, select Try Me > Try Light > Run.

1.3   A tone indicates the reading from the light sensor. If you point the sensor at a light source, such as a window, the frequency will increase.

1.4   Try holding a TV IR remote about five inches from the light sensor and press some buttons. The NXT will chirp immediately. It's like pointing a flashlight at the sensor.

The light sensor is used to perform a variety of functions. By pointing the light sensor down, the robot can follow a black line. Sometimes the sensor is used to prevent a robot from driving off the edge of a table, since the sensor values decrease significantly when an object (such as a floor) is farther away (far objects do not reflect as much light as near objects). The light sensor can also distinguish dark objects from light objects, since dark objects reflect less light.

Light sensors have two modes:

- Active mode—the light sensor LED is illuminated. This is often used for line following or object detection.
- Passive mode—the light sensor LED is extinguished. This mode is used for ambient light detection, such as measuring the sun's brightness.

Figure 1-13 Sound sensor

Sound Sensor

The sound sensor is a brand new addition to MINDSTORMS (see Figure 1-13). Although it resembles a microphone, it really just measures the loudness of sound in decibels (dB). You can't use the sound sensor to record sound files to the NXT brick. Since sound is louder when the source is near, the sound sensor allows robots to home in on sound sources. It can also react to sounds, such as clapping.

Ultrasonic Sensor

Even though the ultrasonic sensor looks like a pair of eyes, it actually has more in common with the sound sensor than a camera. The ultrasonic sensor sends out a sound signal that is inaudible to humans (like a bat), then measures how long it takes for the reflection to return. Since it knows the speed of sound, it can easily calculate the distance the signal traveled.

The ultrasonic sensor is the only I²C sensor included in the NXT kit. It measures distances to solid objects in centimeters or inches. The sensor is capable of measuring distances up to 255 centimeters, though returns are inconsistent at these distances, probably because the return ping becomes weaker. The sensor is accurate from 6 to 180 centimeters, with objects beyond 180 centimeters not reliably located. It has an accuracy of plus or minus three centimeters, though the accuracy is better for close objects.

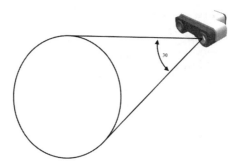

Figure 1-14 The 30 degree ultrasonic sensor cone

The ultrasonic sensor produces a sonar cone, which means it detects objects in front of it within a cone shape. This cone opens at an angle of about 30 degrees (see Figure 1-14). This means that at a distance of 180 centimeters the cone is about 90 centimeters in diameter. The cone shape is ideal for robots, since it is better to scan a large area in front of the robot for possible collisions.

TRY IT: Theoretically a soft object returns less of a sonar ping than a solid object. Let's see how this works in practice.

1.1 *Find a pillow and a solid object, like a book.*

1.2 *Place the objects next to each other at equal distances from the LEGO NXT.*

1.3 *Plug the ultrasonic sensor into port 1.*

1.4 *From the NXT brick main menu, select* View > Ultrasonic (inches or cm) > Port 1.

1.5 *Watch carefully for differences between the pillow and the book. In my tests, it seems the surface of an object does not affect the distance reading from the ultrasonic sensor, however the solid object is detected at longer distances.*

LEGO Building Parts

LEGO has totally revamped the parts included with the NXT kit. There are 577 parts in the NXT kit, which is less than the 718 in the RIS 2.0 kit. The good news is the parts selection is quite good and your desired robot might even be built using fewer parts.

Because some parts previously included are missing, it may be more difficult to build everything you want with just the base kit. Some of the parts missing include tank treads, pulleys, certain gears and the differential. The old RIS kit contained 18 wheels of various types, but this time there are only four wheels and they are identical. This is a significant hindrance if you wanted to build a specialized robot with six wheels, such as rocker-bogie suspension. We'll look more closely at the parts in your NXT kit in Chapter 5.

Using the LEGO Software

LEGO calls their software development language NXT-G. It is very advanced and capable of much more than the previous MINDSTORMS programming environment. The G stands for Graphical, which means you will be using a graphical interface to develop code.

LEGO is not known for software expertise, and the original MINDSTORMS software was somewhat unstable. This time, LEGO turned to Texas-based software developer National Instruments to develop their programming environment. The results are quite amazing.

As with the original software, users create programs by dragging building blocks into an open area from a reservoir of predefined blocks (see Figure 1-15). Each block is essentially a method, and the user selects different parameters for the method using radio buttons, sliders and drop down menus.

The programming environment is more full-featured than might be expected. You can create custom blocks (like methods) that contain lots of code for a specific task. This makes it easier to fit a lot of code into the limited graphical area. The software also allows import of new blocks as new devices become available for the NXT brick.

Although the graphical paradigm is easy to get into, especially for new programmers, it can be hard to manage code beyond a certain size and complexity. If you are used to Java or some other language, it isn't always obvious what a program is doing or how the blocks function.

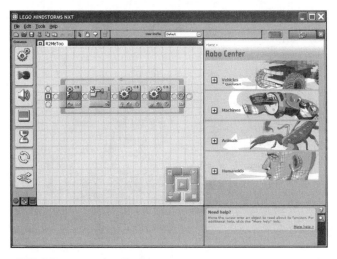

Figure 1-15 NXT-G Programming Environment

Kits and Accessories

The NXT kit provides the basics to create a large number of interesting robots, but you can take your projects even further by expanding the number of parts at your disposal.

Education NXT Base Set – 9797

LEGO Education sells a kit tailored to the educational market. It contains a rechargeable battery and adapter, an extra touch sensor, three RCX conversion cables, additional smaller tires, pulleys and some electric light-bricks. However, there are only 431 parts in total and it costs approximately $30 more than the retail kit.

Education Resource Set - 9648

The education resource set contains a huge assortment of 671 parts. Perhaps most notable are the tank-treads and the differential gear, which are missing from the retail set. Most of the other parts are already included in the retail kit, but there are more of them.

Third party sensors

Sensors are perhaps the most exciting aspect of expanding your LEGO arsenal. These sensors give your robot even greater abilities to sense environmental conditions.

HiTechnic

HiTechnic was the first third-party company to make a deal with LEGO for sensor distribution. LEGO supplies HiTechnic with plastic shells for their sensors, and sells the sensors from the LEGO website. The shells make HiTechnic sensors the most sought after type of sensor since they fit in well with the other LEGO MINDSTORMS parts (see Figure 1-16).

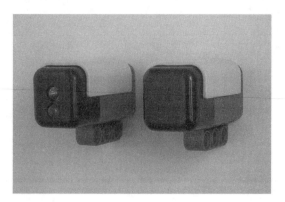

Figure 1-16 HiTechnic sensors

HiTechnic sensors use a black white and grey color scheme as opposed to LEGO's orange white and grey. Because many of their sensors look the same externally (compass and tilt sensors), their sensors are labeled on the bottom.

HiTechnic offers a range of products, including:

- Compass Sensor
- Color Sensor
- Tilt Sensor
- Cables
- RCX Adapter cables
- IR Link with RCX
- IR Seeker
- Port Expander
- Motor Multiplexer
- NXT Prototype board

WEB SITE: *For more information visit:* www.hitechnic.com

Mindsensors.com

Mindsensors.com sells a wide range of sensors. The packaging is not as refined as HiTechnic and they lack a plastic shell, but the sensors are slightly cheaper. The sensors are attached to the robot using LEGO pins (see Figure 1-17). Mindsensors.com includes a complimentary ribbon cable with each sensor. Their roster includes:

- RCX to NXT Communication Adapter
- Tilt Sensor
- RCX Motor Multiplexer
- Compass
- Sony PS2 Controller interface
- Pneumatic Pressure Sensor
- Infrared Distance Sensor
- Temperature Sensor
- RCX to NXT port adapter
- NXT Motor Multiplexer

Figure 1-17 Connecting Mindsensors.com sensors to LEGO

Figure 1-18 Assorted Vernier sensors

Vernier

Vernier sensors are not made specifically for LEGO. However, they offer a special adapter that allows you to plug the sensors into your NXT brick. Be warned that some of their sensors can be costly. They offer more than 40 unique sensors (see Figure 1-18) including:

- pH probe
- Low-g Accelerometer
- Conductivity Probe
- Dissolved oxygen Sensor
- UV Sensor
- Temperature Probe
- Magnetic Field Sensor

WEB SITE: *For more information visit:* www.vernier.com/nxt/

Surveying the Competition

We'll conclude this chapter by looking at some of the other robot kits in the market. Although LEGO dominates the field of do it yourself robotics, there are other competitors. This section will examine the major competition to LEGO.

VEX kit

VEX is LEGO MINDSTORMS strongest competitor. It was originally produced and invented by Radio Shack and later sold to Revell, makers of plastic models. The main programmable unit is reminiscent of the RCX, with 32 KB of programmable memory and a remote controller. The main unit contains an amazing 16 input ports and 8 output ports, however the sensors are basic touch switches and not nearly as imaginative as MINDSTORMS sensors.

Figure 1-19 A VEX robot from the FIRST Vex Competition (photo © 2007 Innovation First)

Unlike the friendly plastic LEGO parts, the VEX kit contains mostly metal pieces with plastic for gears and wheels (see Figure 1-19). Parts attach using regular nuts and bolts. It comes with three analog motors and a servo motor. There are even kits available for pneumatic control.

VEX is more complex but less refined than MINDSTORMS. The FIRST competition (For Inspiration and Recognition of Science and Technology) indicates how MINDSTORMS compares to VEX. The competition has three categories: the LEGO competition, the VEX competition, and the original FIRST Robotics competition. The LEGO competition is the easiest, VEX is moderate, and the non-kit competition is the hardest.

NOTE: *Mindsensors.com offers some sensors for VEX robotics.*

WEB SITE: *For more information visit* www.vexlabs.com

fischertechnik ROBO Mobile Set

The fischertechnik set is often seen as the closest competitor to LEGO. They've been around since 1965; longer than the LEGO Technic line of parts. They introduced their first computer programmable kits in the late eighties, but they've never experienced the same level of success as LEGO.

The basic ROBO Mobile Set contains two motors, four sensors, and more than 350 parts for $320. The main programmable unit, the ROBO Interface, contains 128 KB of flash memory, a USB connector, and a 16-bit microcontroller. It contains four output ports for motors and 14 input ports of different types (digital and analog or varying voltages).

Figure 1-20 A typical fischertechnik robot (photo copyright 2007 fischertechnik)

The fischertechnik kits are generally meant for a slightly more advanced crowd than LEGO, and seek to teach mechanical engineering concepts (see Figure 1-20). Some of the kits, such as the Industry Robots kit, are very interesting.

 WEB SITE: *For more information visit* www.fischertechnik.com/html/ computing-robot-kits.html

Handyboard

The Handyboard was invented by Fred G. Martin, the same engineer who designed the original MIT programmable brick (see Figure 1-1). It is a freely available public domain design, but several manufacturers produce and sell the Handyboard (see Figure 1-21). The kit requires knowledge of electronics to assemble, so it's for the soldering crowd only.

Figure 1-21 A Fully Expanded Handyboard

Handyboard uses an 8-bit Motorola 6800 derivative called the 68HC11 processor. Handyboard contains 32 KB of memory, an LCD display, four motor ports, and eight digital and analog input ports. Programmers use Interactive C to program the kit. Handyboard has been around since before the RCX and it is starting to show its age when compared to the latest generation of kits.

NOTE: *Handyboard 2 is slated for release in 2007.*

WEB SITE: *For more information visit* www.handyboard.com

leJOS NXJ

Topics in this Chapter

- Introducing leJOS NXJ
- Installation
- Your First leJOS NXJ Program
- Setting up a Development Environment
- R2-MeToo

Chapter 2

L EGO's NXT-G programming language is great, especially for those just learning to program. However, if you are already a programmer and feel more comfortable with an enterprise-level programming language or if you think you have a future in programming, you may want to move to a more advanced programming language.

This book uses leJOS NXJ, a software package that allows you to program the NXT brick using Java. Why choose leJOS NXJ over the NXT software? There are many answers but the most important is that leJOS NXJ allows you to do things you can't do with the standard LEGO software. For example, the mapping project in chapter 13 plots real-time data on your PC from the NXT. There is just no way to do this with the standard LEGO Software. There are also some very complex navigation classes in NXJ which would be difficult to program in NXT-G. I'm not saying NXT-G can't do this - it probably can. It just gets pretty complicated, graphically, when the programming gets this complex.

If you are a Java programmer, you probably already know just how good the language is. This chapter will introduce what Java programmers have been craving – a Java environment complete with threads, arrays, floating point numbers, recursion, and total control of the NXT brick. This chapter will also show how to set up leJOS NXJ on your computer, and just as important, a time saving Integrated Development Environment (IDE).

Introducing leJOS NXJ

The leJOS NXJ platform is an extension of leJOS for the RCX brick. The leJOS developers took the latest version of the leJOS RCX code and ported it to the NXT. Rather than keeping the code backward-compatible with the RCX, we decided to create a separate branch of code specifically for the NXT.

NOTE: *The name lejos means 'far' in Spanish. In the name leJOS, the letters JOS are capitalized because those letters stand for Java Operating System. Since le means 'the' in several languages, this would mean the Java Operating System. NXJ refers to the part of the package that is specifically for the NXT brick. This includes Java classes such as lejos.nxt.Motor (which will be covered in chapter 4). NXT is a registered trademark, so we use NXJ to signify a link with Java while still indicating NXT compatibility.*

The leJOS JVM is written in C code in a very platform independent style, which means it is easily ported to other machines. So far it has appeared on the RCX brick, the Gameboy Advance, and the NXT brick.

There are also tools on the PC side to compile and upload code to the leJOS JVM. leJOS is multiplatform, and these days that means Windows, Linux, and Macintosh. leJOS NXJ is available for each of these platforms, allowing you to develop NXJ code under your favorite operating system.

Let's install leJOS NXJ and see what it can do.

Installing NXJ

You receive the following when you download the Java SDK from Sun Microsystems:

- classes that allow you to interact with your computer
- tools to compile code
- a Java Virtual Machine to run your code

You get the same things when you download leJOS NXJ, except that the class selection is tailored to the capacity and needs of the NXT brick. Let's install these tools.

NOTE: *The leJOS download contains the latest setup instructions. If you experience any difficult using the steps below, refer to the instructions as some procedures might have changed.*

Windows

1. Your first step on the path to leJOS bliss is to download the latest version from www.lejos.org.
2. Unzip the contents into a directory. e.g. c:\java\lejos_nxj
3. Now we need to set some environment variables. Select Start > Control Panel > System. Click the Advanced tab, then Environment Variables (Figure 2-1).

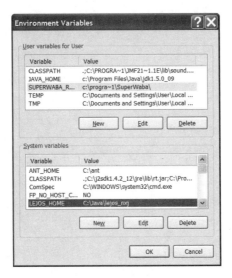

Figure 2-1 Viewing the environment variables

4. Click New to create a new environment variable. It can either be in System variables (if multiple users will use leJOS) or User variables if your account is the only one using leJOS. Type LEJOS_HOME as the variable name and add the leJOS directory (see Figure 2-2). Click OK.

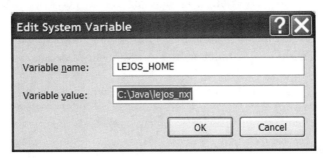

Figure 2-2 Setting the leJOS home directory

5. While you are in the environment variables, check to make sure JAVA_HOME has been set up. If not, add this variable and type in the directory to your Java SDK.

6. Finally, add the bin directory to your path so you can use the leJOS compiler tools from the command line. Add the following to the end of your path variable (see Figure 2-3).

```
;%LEJOS_HOME%\bin
```

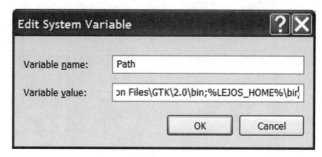

Figure 2-3 Setting the path to the leJOS binaries

That's all. Make sure you have already installed the LEGO NXT software, as its USB drivers are used. You can now skip down to Uploading Firmware.

Linux

Before getting started, download and install the latest version of the Java Development Kit (JDK) from Sun Microsystems. Your PATH must contain the JDK's bin directory. Also, make sure you have set the JAVA_HOME properly to the root directory of the JDK.

1. Download and decompress the tar file from www.lejos.org.
2. Set your environment variable LEJOS_HOME to the directory you installed leJOS.
3. Add the leJOS bin directory to your PATH. Depending on the privilege settings, you might need to adjust the execution permissions in the bin folder.
4. Your PATH must also contain the ant binary (ant 1.6 or above).
5. You need libusb installed so the leJOS tools can access your USB port.
6. Now you need to build the distribution. Switch to the build folder and run ant. Note that depending on the privilege settings you might need to adjust the execution permissions in the release folder.

That's all. You can now skip down to Uploading Firmware.

Mac OS X

Macintosh owners can download binary files compiled just for their system. These files are a universal build, meaning they will work on both Power PC and Intel based Mac OS X computers.

The leJOS tools for compiling and uploading Java code run in a shell environment, such as tcsh. Before you can do that, you will need to set up some environment variables for the tcsh shell.

1. Download the Mac OS X distribution from www.lejos.org.
2. Extract this file into a new location, such as /Applications/lejos_nxj.
3. If you use the administrator login, you will need to create (or edit) the file *.tcshrc* in your user home directory. Run textedit from your Applications folder.

4. Type the following two lines into the window (using the directory where you extracted leJOS), then save:

```
setenv LEJOS_HOME /Applications/lejos_nxj
setenv PATH ${PATH}:${LEJOS_HOME}/bin
```

5. Select your administrator directory (/users/administrator), and type *.tcshrc* as the file name. Uncheck the box saying "If no extension is provided, use '.txt'" before you save. Then you'll get a warning box suggesting these names are reserved for the system (see Figure 2-4). Click Use '.' and it will save.

NOTE: *If you prefer to use csh instead of tcsh, you should instead edit/create the file .cshrc with the same lines.*

6. Bring up a Terminal window and type *tcsh*. You'll now be in a tcsh shell.

7. Type *setenv* to make sure your PATH and LEJOS_HOME variables are set up correctly. That's all! You are ready to test leJOS.

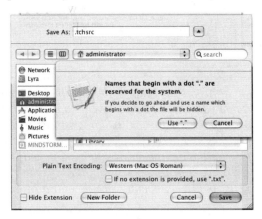

Figure 2-4 Saving the environment variables in Max OS X

Uploading the Firmware

It is much easier and faster to upload firmware to the NXT than it was with the RCX brick.

1. First we need to place the NXT brick in firmware upload mode. The reset button is cleverly hidden in a LEGO pin hole so you don't accidentally press it. Turn on the NXT. Using a bent paperclip, insert it into the hole in the upper-right corner of the underside of the NXT brick (see Figure 2-5). Hold the button for at least four seconds to erase the current firmware and put it into firmware upload mode.

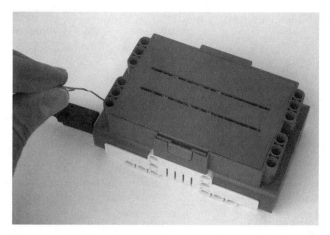

Figure 2-5 Using a paperclip to activate firmware upload mode

2. Your NXT brick should be making a soft pulsing sound. Now we need to upload the firmware. Plug in your USB cable and type:

   `lejosfirmdl`

3. After a very brief moment you will see the leJOS NXJ logo and a menu system will appear. Your NXT brick is now ready to accept Java code.

WARNING: There is a theoretical limit to the number of times you can replace the firmware on your NXT before it wears out. Every time you replace the firmware, a piece of data called a lock bit is used up. This bit is rated to work 100 times (minimum) before it expires. However, don't let this dissuade you from replacing the firmware with something you want more. Most engineers agree that the lock bit will last far in excess of 100 times. Chances are you will never even notice this limitation.

Compiling and Uploading Java Code

In this section you can try compiling and uploading some sample code from the command line. Windows users can enter the command line by selecting Start > Run and then typing cmd (Click OK).

1. From a command line prompt, change to the samples\tune directory where you installed leJOS.

   `cd \java\lejos_nxj\samples\tune`

2. You can optionally open the file Tune.java with a text editor to view some leJOS code. Compile the sample Java file:

   `lejosjc Tune.java`

3. This creates a file called Tune.lej. Now it is time to upload the this file. Plug in your USB cable (if you own a Bluetooth dongle you don't have to plug this in). Turn on the NXT by pressing the orange button and type:

```
lejosdl Tune.lej
```

Select the file from the menu to hear a tune play.

Installing a Development Environment

As we have seen, Java programming is possible with a text editor and a command line. However, it's easier to click on buttons to make things happen rather than typing commands and optional parameters. Also, most text editors don't have many features to help you enter code. It won't tell you when you've misspelled the name of a class or forgotten a bracket.

An IDE, or Integrated Development Environment, is a tool that allows you to enter, compile, and upload code to your NXT using simple buttons. It also monitors code syntax, coloring your code so you can more easily identify the parts. This section will suggest a free, open source IDE for your leJOS NXJ needs.

One of the best open source IDEs is Eclipse by IBM (see Figure 2-6). It's free, powerful, and easy to use. It makes sense to use a more advanced IDE with the NXT since your code can grow quite large.

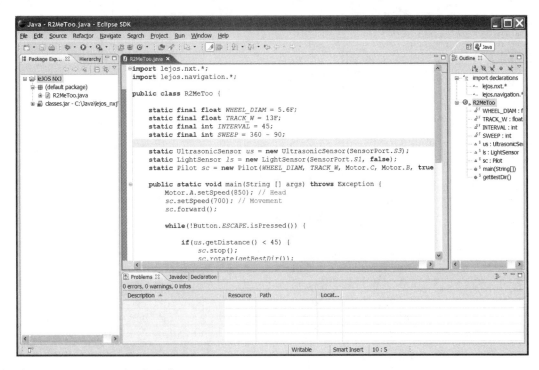

Figure 2-6 Programming in Eclipse

Setting Up Eclipse

1. Download Eclipse from: www.eclipse.org

2. Decompress the files into a directory. This will be the permanent location for Eclipse.

3. That's it. Eclipse uses no setup and doesn't store registry settings or copy native library classes to other directories. To run Eclipse, double click the executable file in the Eclipse directory (or create a shortcut to this). To uninstall Eclipse, merely delete the Eclipse directory from your computer.

When you first run Eclipse you can browse some optional help files and tutorials. If you want to get right to Eclipse programming, close the Welcome tab.

Using Eclipse with leJOS NXJ

Now that you have Eclipse installed it's time to configure it for leJOS NXJ.

1. Set up a project in Eclipse for your leJOS NXJ code. Click on File > New > Project and you will see the new project Wizard (see Figure 2-7). Select Java Project and click next.

Figure 2-7 Creating a new project in Eclipse

2. For project name enter something like leJOS NXJ. Click Next and then Finished. Eclipse creates a new folder for your leJOS NXJ code.

3. Now we need to set the Eclipse classpath to the leJOS classes. In the Eclipse menus select Project > Properties and you will see a Properties window for your project. Select the Libraries tab and select Java Build Path in the left hand frame (see Figure 2-8).

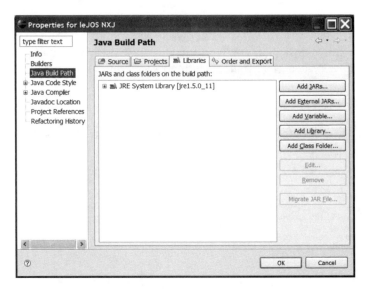

Figure 2-8 Changing project properties

4. Click on Add External JARs... and browse to the classes.jar file in the lib directory where you installed leJOS. Click OK.

5. Now we need to set up the leJOS tools to compile and upload code to your NXT brick. Select Run > External Tools > External Tools... to bring up a new window. Select the Program item and click the New button (see Figure 2-9).

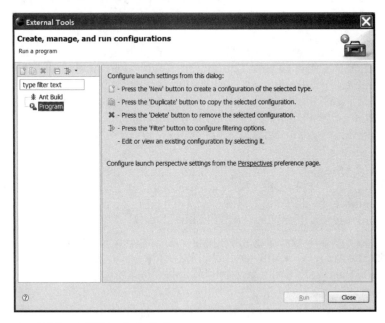

Figure 2-9 Adding leJOS tools to Eclipse

6. First we'll create the compiler tool (see Figure 2-10). Type in leJOS Compile. For location, click Browse File System and browse to the \bin\lejosjc.bat. For Working Directory click Variables and choose project_loc. Under arguments type in the following, then click Apply:

$\{java\_type\_name\}$.java

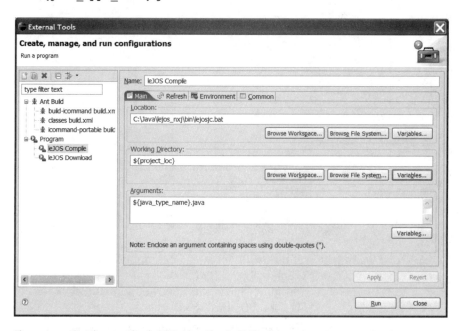

Figure 2-10 Setting up the leJOS compiler in Eclipse

7. Now we'll create the tool used for transferring code to the NXT. Click the New Launch Configuration button again and type in the name leJOS Download. For location browse to bin\lejosdl.bat. For Working Directory click Variables and select project_loc. Finally in arguments type the following and click Apply:

$\{java\_type\_name\}$.lej

8. Select the arrow next to the Tools icon (green arrow with red tool kit), then select Organize Favorites. Click Add and add all three of the leJOS NXJ tools. Click OK. You can now organize them individually by clicking on them and selecting up or down. I prefer the order leJOS Compile, leJOS Link, and leJOS Download. Click OK.

You can now compile and upload code to your NXT. When you select the arrow you will see the tools, ready for action. You can try using Eclipse with the example robot given below.

So what is so good about Eclipse? Try making an error in your code. Notice that right away it tells you where there is a problem? It lists all the errors or warnings, shows you exactly where they are, and even tells you what is wrong and how to remedy the problem. When you type an object name and press the period key it even pops up all the methods you can access.

Although Eclipse is simple, it has many tools to make your coding experience easier. Briefly, click on the Source menu item and look at some of the tools. Cleanup lets you import code and clean it up to your preferred style, or clean up your own code. If your indentations are wrong, highlight your code and select Correct Indentation. Instantly it fixes everything. There are also functions for generating try-catch blocks and Getter/Setter methods.

Sometimes you need to overhaul your code in major ways. Check out the Refactor menu item, which you can use when you want to rename a class. If you want to rename a variable, right click it and select Refactor > Rename. You can even overhaul the architecture of your code, such as making an interface from a class. These methods go through your entire code making changes.

TIP: Eclipse can be enhanced using a variety of plugins. For example, it can be difficult and time consuming to create a graphical user interface (GUI) using raw Java code. You can download a plugin that allows you to graphically design a GUI for your program which then automatically generates the code. Most plugins are free for non-commercial use. Eclipse plugins can be found at: www.eclipseplugincentral.com

Plugins are installed through Eclipse (Help > Software Updates > Find and Install...) *however you need to enter the remote location of the plugin supplier. The remote location is typically located on the web page for the project.*

Beyond Hello World

Let's try a more heavy duty project with leJOS. In the first chapter we showed some NXT-G code that actually controls the robot below. If you want, you can build the robot and use the NXT-G code in Chapter 1 with it. However, in this section we will program the robot in Java.

Building the Robot

Our first project is a simple robot to familiarize you with building and programming. This little android might not be able to traverse the sands of Tatooine but he performs well on living room carpet.

 TIP: The following LEGO plans are easier to understand if you first gather up the parts for the current *step (shown in the upper left corner) and then assemble the step. If you pick parts one at a time out of your box it will take longer and you might lose track of whether you've used the correct number of parts.*

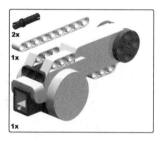

STEP 1 Add parts as shown.

STEP 2 Make sure the axle protrudes by one unit as shown. The small beam will hang free for now.

STEP 3 Make sure the axle protrudes by one unit as shown.

STEP 4 Add parts as shown.

STEP 5 Add parts as shown.

STEP 6 Attach one gear to the axle. The other gear is an idler hear, residing on a yellow axle pin.

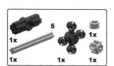

STEP 7 The black angle-connector turns freely for now.

STEP 8 The beam on this side uses a long pin and a short pin as shown. The 3-unit beam on the other side is partially obscured by the topmost gear. This beam attaches to the top of the white beam, and uses a long pin in the middle hole.

STEP 9 The beam on this side uses two short pins. The split-axle joiner on the other side (partially obscured) uses a single long pin in the top hole. It is more visible in step 10.

STEP 10 Line up the black angle-connector from step 7 and place the turntable. Insert the 4-unit axle into the third hole into the angle-connector, and then place a bush on the end.

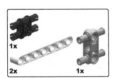

STEP 11 Insert the black 4-pin connector into the two remaining holes of the turntable, and then add the 7-unit beam. On the other side (hidden) insert the grey 4-pin connector into the turntable, and then add the other 7-unit beam.

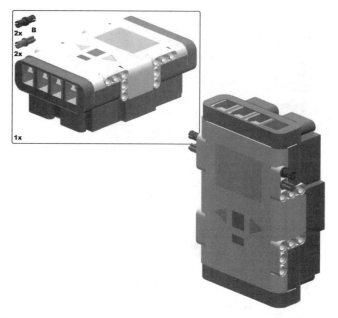

STEP 12 Add parts as shown.

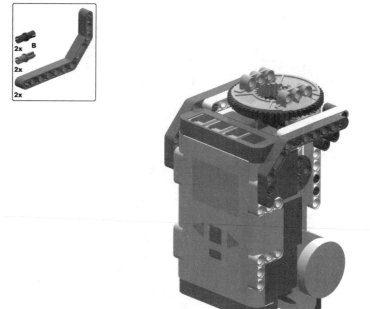

STEP 13 Join the bent beam to the closest side, then place a blue pin in the far end, and a black pin as shown. Join the previous assemblage. Then join the bent beam on the other side.

STEP 14 Add parts as shown.

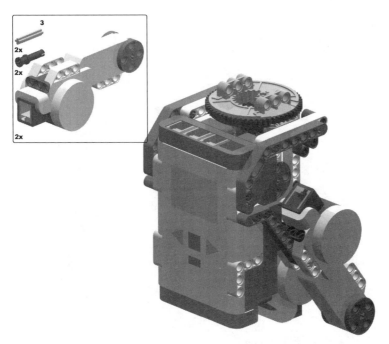

STEP 15 When adding the motors, place the long pin with bush on the inside into the top hole of the motors.

STEP 16 Add parts as shown.

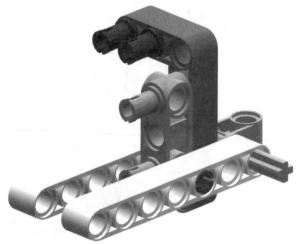

STEP 17 Add parts as shown.

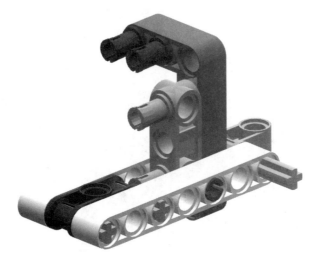

STEP 18 Add parts as shown.

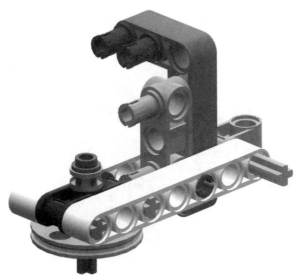

STEP 19 Add parts as shown.

STEP 20 Add parts as shown.

STEP 21 The blue pins join the 3-unit beams to the black beams on the closest side.

STEP 22 Add parts as shown.

STEP 23 Add the wheel to the base of the robot as shown. The blue pins go into the cross-hole of the L-beam (hidden).

STEP 24 Insert the axles into the motors. The half-bush goes on the axle first, then the tire.

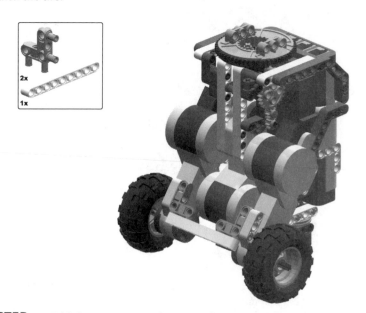

STEP 25 Add the rear support beam as shown.

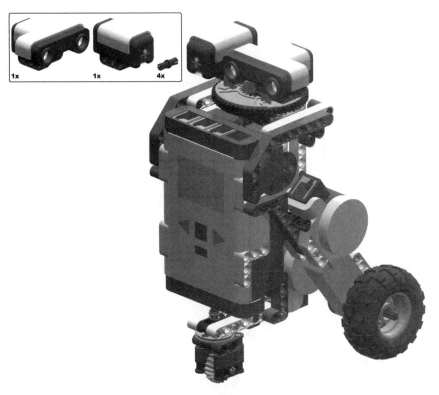

STEP 26 Add the sensors, one on each side of the turntable. You're done!

Now plug three cables into ports A, B, and C. Motor B (right leg) uses a longer cable, while the rest use medium cables. The cables go under the turntable so they won't block the ultrasonic sensor. Feed each cable out the back, under and around the lower support on the back, then up to each motor.

Use a medium cable to connect the ultrasonic sensor to port 1. If this is not done correctly it can hinder head movement. Plug the cable into the NXT first and then feed it up through the center of the body and through the hole in the middle of the turntable. Make sure the slack in the cable is at the top of the robot, then plug in the ultrasonic sensor. (The light sensor is unused in this project, but if you want you can connect it to turn on a decorative light.)

Entering Code

The robot will wander around the room until it gets too close to an object. Then it will stop and figure out a new direction to travel. To enter the code in Eclipse, create a new class in your project by selecting File > New > Class (see Figure 2-11). Type the class name as R2MeToo and click Finish. If you are using command line, open a text editor and save the file to a directory. You will need to create a directory to run your code from.

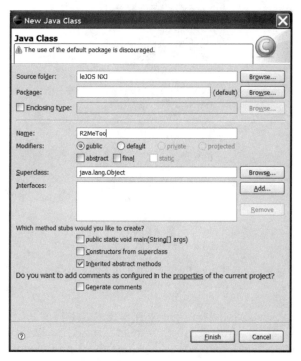

Figure 2-11 Starting a new class in Eclipse

 NOTE: *I strongly recommend downloading all the code in this book straight from the book's website. This code is continually updated in case leJOS NXJ changes and it saves you from typing the entire code yourself. The website is at:* www.variantpress.com

 NOTE: *Do not type in the numbers in front of every line. These are for reference purposes only.*

```
1. import lejos.nxt.*;
2. import lejos.navigation.*;
3.
4. public class R2MeToo {
```

```
5.
6.    static final float WHEEL_DIAM = 5.6F;
7.    static final float TRACK_W = 13F;
8.    static final int INTERVAL = 45;
9.    static final int SWEEP = 360 - 90;
10.
11.   static UltrasonicSensor us = new UltrasonicSensor
          (SensorPort.S3);
12.   static LightSensor ls = new LightSensor(SensorPort.S1,
          true);
13.   static Pilot sc = new Pilot(WHEEL_DIAM, TRACK_W,
          Motor.C, Motor.B, true);
14.
15.   public static void main(String [] args) throws
          Exception {
16.     Motor.A.setSpeed(850); // Head
17.     sc.setSpeed(700); // Movement
18.     sc.forward();
19.
20.     while(!Button.ESCAPE.isPressed()) {
21.
22.       if(us.getDistance() < 45) {
23.         sc.stop();
24.         Sound.twoBeeps();
25.         sc.rotate(getBestDir());
26.         sc.forward();
27.       }
28.       Thread.sleep(200);
29.     }
30.   }
31.
32.   // Rotate head and find longest direction
33.   public static int getBestDir() {
34.     ls.setFloodlight(true);
35.     int bestDir = 0;
36.     int bestDist = 0;
37.     for(int i=-SWEEP/2;i<SWEEP/2;i = i + INTERVAL) {
38.       Motor.A.rotateTo(i * 9); // 9 = gear ratio
39.       int curDist = us.getDistance();
40.       if(curDist > bestDist) bestDir = i;
41.       if(curDist >  200) break;
42.
43.     }
44.     Motor.A.rotateTo(0);
45.     ls.setFloodlight(false);
46.     return bestDir;
47.   }
48. }
```

Compiling and Uploading in Eclipse

1. Enter the R2MeToo code above into the document and save it.
2. Try compiling R2MeToo from the Tools button in the Eclipse toolbar. When you compile, you probably won't see any output in the console area unless there are errors.
3. Now turn on your NXT brick, plug in the USB cable, and select the leJOS Download tool.

If you want to compile from the command line without Eclipse, use the commands below.

1. From a command line prompt, change to your code directory.

   ```
   cd \java\nxj_code
   ```
2. Compile the Java file:

   ```
   lejosjc R2MeToo.java
   ```
3. Now upload the file to the NXT. Plug in your USB cable, turn on the NXT by pressing the orange button and type:

   ```
   lejosdl R2MeToo.lej
   ```

Running the Program

Set your robot down on a smooth floor and make sure the head is centered. If it is off center, pull one of the gears off the robot and rotate the head until it is straight, then put the gear back on. R2MeToo is very good at finding his way around, though low-lying objects might give him some problems because his sonar is so high. Because his "eyes" don't require light he can operate in pitch black, much like a bat. R2MeToo looks sleek running in the dark with the floodlight turned on.

TIP: You can stop code on the NXT by pressing the orange enter button and grey escape button at the same time. This is much like using Ctrl-Alt-Delete to end programs in Windows.

The Java Virtual Machine

As previously mentioned, the Java VM runs your Java code. A JVM is like a virtual computer. This allows the same code to run on Windows, Linux, Macintosh or other platforms. The leJOS JVM has virtually everything the standard Sun JVM has. To describe the leJOS Java VM is to describe Java. Let's go over the basics.

Memory

The RCX had 32 KB of memory, with the leJOS JVM occupying approximately 16 KB, leaving another 16 KB free for user programs. This meant most projects were fairly minimal. With the memory available on the NXT, you can make a high IQ robot. There's even enough memory to include hidden Easter egg functions in your programs. You can also play recorded audio. In short, your robot can now become much more interesting, like the difference between having a conversation with Marcel Marceau and Philip K. Dick.

The leJOS JVM is stored in about 27 KB of memory, leaving you with at least 229 kb for user programs and data (about 15 times more free memory). A troublesome barrier has been negated.

Speed

Speed is usually not a great issue for robotics. The leJOS RCX was fast enough even for time-sensitive applications like balancing on two wheels. It's worth mentioning that leJOS NXJ is faster than the previous version. In speed comparisons, it took leJOS NXJ approximately 250 milliseconds to perform the same calculations that took leJOS RCX about 1500 milliseconds. That's speed increased by a factor of six.

Floating Point Numbers

The leJOS JVM allows floating point numbers (decimal places). This gives leJOS the ability to represent fractional numbers, and it allows trigonometry functions such as tan, cos, and sin. Without floating point numbers it is not very practical to perform the trigonometry functions vital for navigation.

WARNING: *leJOS tries to save memory by not fully supporting 64 bit types, such as double and long. You can do anything with doubles in leJOS, including initializing double variables and performing mathematical operations on them. However, internally they are treated as floats. This means that although it looks like you are using a 64 bit double number, in fact it has the same accuracy as a 32 bit float.*

Threads

Multi-threading allows different parts of a program (threads) to execute at roughly the same time, sharing the CPU. The leJOS thread scheme is very complete, allowing synchronization and interrupting. Since there is a single processor inside the NXJ (as opposed to computers containing multiple processors) it is necessary for the threads to take turns using the processor in a pre-emptive scheme.

Threads are usually controlled by a thread scheduler, which allows a thread to take control of the processor for a number of instructions. leJOS threads are handled by a very simple scheduler. It keeps a list of current threads and simply switches from one thread to the next using a single native C function called switch_thread(). The scheduler does this all automatically. In the latest incarnation of leJOS, each thread is allowed to execute up to 128 instructions before the next thread is given a turn.

leJOS theoretically allows up to 255 threads to be created! This is of course more than enough threads for most robotics projects, and illustrates the few limitations imposed by leJOS. Keep in mind, however, that every thread you create uses memory.

Arrays

Arrays are useful for storing large sequences of related numbers or objects. Even multi-dimensional arrays are allowed in leJOS. They are potentially useful for navigation systems that keep track of location using a two-dimensional grid based system. Larger dimensions can also be used, but (without getting into technical detail) not too many, otherwise the JVM might run out of resources and crash. Programmers rarely use more than two dimensions, however, so this is not a large factor.

Event Model

leJOS uses the Java event model, which includes listeners and events. The Button and SensorPort classes use this model, as demonstrated in Chapter 4. This model allows for clean, easy to understand code when the NXT waits for an event to occur.

In the event model, an object acts as an event source, such as a sensor (imagine a touch sensor). One or more objects can register with the sensor to listen for events. When an event occurs (such as the touch sensor being pressed) all listeners are notified of the event and can respond accordingly (Figure 2-12).

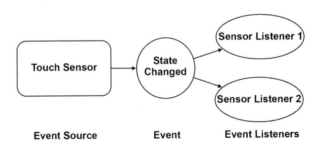

Figure 2–12 Touch Sensor Event Sequence

Exceptions

Java uses an error handling scheme to deal with errors (exceptions) when they occur. The advantage of exception handling is it allows error checking code to be separated from the regular program logic. This makes the code tidy and easy to understand, rather than forming a tangled mess of error checking code and logical code. The common term for this is exception handling.

The main virtue of exceptions is that they propagate up the call chain, so it is not necessary to write methods with return values to indicate error codes. For example, exceptions allow you to seamlessly distinguish between a legitimate null return value and a method error.

Additionally, Java programs give stack traces instead of crashing the system, which is useful for locating the source of the error.

Recursion

Recursion is a programming technique that allows a method to call itself. Java, and most other object oriented languages, allows recursion. leJOS is limited in the number of times a block of code can enter itself, however. Currently it can only go about 10 levels deep, but if the method is using many local variables then this number will decrease.

Garbage Collection

Object oriented languages such as Java rely on the creation of new objects in order to function. When an object is no longer in use by a program it should be removed from memory, a process known as garbage collection. Most Java implementations have automatic garbage collection, but leJOS does not as of this writing. This might sound like a problem, but is not a big factor. It just requires a programmer to be mindful of reusing objects rather than using the keyword *new* to create new ones. You probably won't notice the absence of a garbage collector unless you are creating lots of temporary objects in your code.

NOTE: *Primitive data types such as int and float are not objects, and do not need to be collected when the program is finished with a variable.*

Exploring the NXJ Platform

The leJOS API includes the package lejos.nxt just for controlling the NXT. This section is a brief overview of what you can expect from the NXJ classes. (Chapter four will explore how to use the lejos.nxt package.)

Motors are Everything

Every device has a primary function; the main reason you use that device. When you boil away everything else, the main reason for owning an NXT is to control motors. Even the sensors are there to serve the motors. So leJOS NXJ lives or dies depending on how well it can control motors. This is where leJOS has an advantage over the LEGO firmware, in my opinion.

With LEGO firmware, if you tell the motor to rotate forward 40 degrees, it rotates, slows down, overshoots the mark, then backs up and finally stops within five degrees. With many projects, this is not ideal. We wanted to improve the function so that it decisively moved to the destination, and when it got there, stopped without overshooting.

leJOS gives you a very refined algorithm that took a lot of time to perfect. The person behind these algorithms is Professor Roger Glassey of Berkely. His algorithm rotates the exact number of degrees, stopping exactly at the desired number. Whether you request a 45 degree rotation in leJOS directly on the brick or via Bluetooth, you will receive accuracy.

You can also control the motors by speed. You tell it the speed to maintain and the program will monitor the motors and change the power level to maintain a constant speed. If a robot is going up a hill or down a hill, it will maintain the same speed. You can even check to see how fast the motor is rotating.

Sensors

It was easy to develop sensor code for the RCX. They all used the same basic Sensor class because they expressed their measurements as values between 0 and 1023 (even the touch sensor). The NXT sensors are far more complex and they provide much different output. We decided, therefore, to write unique classes for each NXT sensor. For you, this makes using sensors in leJOS NXJ far simpler than before.

Buttons

All four buttons on the NXT can be reprogrammed under leJOS NXJ. You can even use events to listen for button presses and react accordingly when one is activated. This makes it easy to separate the user interface portion of your code in an object-oriented style.

System Time

Time is kept on the NXT brick as the number of milliseconds that have elapsed since it was turned on. Think of this as a time sensor for your robot. This can be useful for time-stamping when events have occurred.

Battery Power

leJOS is capable of checking battery charge. You might think this would give you a percentage of battery power remaining, but it actually returns a number representing the voltage of the batteries. If the battery charge starts to fall rapidly it usually means the NXT is "running on empty".

LCD Display

leJOS NXJ allows you to take complete control of the LCD. You can draw lines, shapes and bitmap images to the display. leJOS even includes a complete character set so you can output text and numbers.

Sound

The leJOS API allows you to play simple sounds. Using iCommand (Chapter 7), you can upload and play prerecorded sound files.

Bluetooth Communications

Communications can take place between the NXT brick and any other Bluetooth device. Not only can you communicate with other NXT bricks via your PC, but you can also communicate with mobile phones, GPS units, PDAs, and other devices.

Robotics Classes

The leJOS developers believe that if a class is useful for general robotics tasks, it should be part of leJOS. For example, navigation is a big part of robotics so the leJOS NXJ platform includes a variety of packages dealing with navigation.

iCommand contains a package for vision based robotics using a video camera (Chapter 10). It also has a package for subsumption architecture (Chapter 18). These classes allow you to jump over the low-level robotics programming tasks and go straight to programming more interesting behavior.

As of this writing, leJOS is the only Java option for the NXT brick. If you like Java, this is the environment for you. The developers think it holds up well against other platforms for the NXT, such as C language or the standard LEGO software. However, don't hesitate to try other languages for the NXT brick, such as NXC. C language is important in the programming world, and often goes hand in hand with Java.

Java for Primates

Topics in this Chapter

- Java Fundamentals
- Core Java Language
- java.lang

Chapter 3

T he Dummies series of books teaches absolute beginners. This chapter, Java for Primates, attempts to one-up those books by simplifying things even more and possibly attracting a new audience. If this is your first time programming Java, you are in for a rewarding experience. Java has consistently been the most popular programming language out there, beating both C and C++ year after year. Java is a popular and well loved language for a reason.

Although this chapter will not teach you everything you need to know about Java, it will teach you enough to program your NXT brick using leJOS. Everything you learn here also applies to the standard Sun Java language. After all, leJOS is Java; it just contains a smaller set of classes. The core language is identical in syntax to standard Java. After becoming proficient in leJOS you will find it easy to make the leap to programming Java on any computer.

 NOTE: *This chapter is an introduction to Java. There are many good books on the market that teach the Java language in greater depth, such as* Core Java *by Cay Horstmann.*

For those who already know Java, I suggest skipping over most of the section titled Java Fundamentals. This section deals with the absolute basics such as flow control, how to make methods and classes, and topics familiar to seasoned Java programmers. You might want to skim the Notes and Warnings, since they highlight the differences between leJOS and standard Java.

Java Fundamentals

Every programming language uses keywords and symbols to provide core functionality. This functionality includes such things as declaring classes, methods, and variables. It also includes manipulation, such as adding two variables. A programming language would be boring if it could only run commands one after another, so to give a program a branching behavior you need to be able to control program flow—in Java by using *if, for, while* and *do* statements. In this section we will cover each of these fundamentals and show examples in actual Java code. Before getting into Java it would be a good idea to touch on a concept fundamental to Java: Object Oriented Programming.

OOP

Object oriented programming (OOP) is a simple but powerful concept. In order to appreciate OOP it should be compared to structured programming. Structured programming uses one body of code with multiple methods. This type of program is difficult to manage because methods and global variables are constantly added to the same program. As a program grows larger it becomes difficult to manage and error prone.

OOP has a significant difference that has powerful implications: code can be segregated into discrete units. These units, or objects, are usually defined by a theme. For example, imagine you are trying to program a robot to compete in a Chess tournament. The robot needs to be able to move pieces, and also to play chess. One object would contain functionality centered on the theme of piece moving, and another object would contain functionality for playing chess (Figure 3-1). These units are defined by classes in Java. The RobotArm class would contain all the methods to move chess pieces around the chess board, and the ChessPlayer class would contain all the methods for choosing which piece to move. Most likely the ChessPlayer class would call methods from the RobotArm class in order to function.

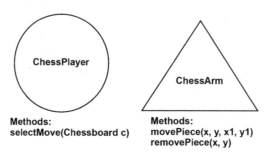

Figure 3-1 Objects in a robotics application

Classes are created using methods (functionality) and fields (data). Once a class is defined, it can be *instantiated*, which means an object is created which can be manipulated in code. One class can be used to create many objects. Another powerful feature of OOP is *type extensibility*. This is the ability of classes to inherit data and functionality from other classes. These topics will be fleshed out with actual examples in the following sections.

Source Files

In Chapter Two you were shown how to set up your development environment and compile a simple program for the NXT. Let's now analyze the contents of a Java source file:

```
1. import lejos.nxt.*;
2. class Hello {
3.    public static void main(String [] args) {
4.       LCD.drawString("HELLO", 2, 0);
5.       LCD.refresh();
6.       while(true){}
7.    }
8. }
```

Enter this code in a file named Hello.java and compile it. Once it is compiled, upload it to the NXT and run it (refer to Chapter Two if you are not familiar with how to do this on your system). You should see the word HELLO appear on the LCD display. In order to escape from this loop, press the orange and dark grey buttons at the same time. The program is uninspiring, but it demonstrates some basic concepts.

This is about as simple as a leJOS Java source file can be. For now, we'll ignore the import statement in the first line. The second line contains a statement that defines the class name, in this example, Hello. Java is case-sensitive which means it notices if letters are capitalized. To a Java compiler, the word "Hello" is different from "hello". Note the curly brace located after the class declaration and the matching curly brace located at the end of the file. All of the methods and variables contained in the class must lie within these two curly braces, otherwise it will not compile.

Line three contains a method definition. As you can see, methods also use curly braces. This method is named main(); it is the main method that starts the ball rolling. All leJOS programs are started using a main() method. There are other keywords in the method definition: public, static, and void. For now, the only thing you should know is that these must be present for the main() method to function properly. There are also some words in parentheses after the method name, the significance of which will be explained further in the chapter.

The statements within the main() method are the heart of the program; they provide the functionality of the class. The first line outputs the word "HELLO" to the LCD display, and the next line puts the program in an endless loop. Without this endless loop the program will flash "HELLO" for a brief millisecond then go back to the regular display.

Classes

Classes are programming structures that define objects, sometimes called *meta-objects*. They are like templates that are used to create objects, typically by using the *new* keyword. One class can make a multitude of objects, much like a rubber stamp can create copies of the same design. For example, String is a class contained in the java. lang package. In order to create several String objects, we can use the following code:

```
String s1 = new String("String A");
String s2 = new String("String B");
String s3 = new String("String C");
```

Once an instance of an object is created it is possible to call methods on that object or access variables. The following line of code uses the toCharArray() method to retrieve a character array of the String object created above:

```
char [] name = s1.toCharArray();
```

This is the essence of object oriented programming. Objects contain all the data and methods they require in one place. This makes the code easier to understand, especially when compared to structured "spaghetti" code where methods and data are thrown together.

Until now we have seen only some very basic class definitions, but classes can also be modified by a number of keywords. For example, the following class contains a declaration using many keywords:

```
public abstract class Hermes extends Navigator implements
SensorListener {}
```

Let's examine each of the keywords available for class definitions.

Class Access

Classes can be either public or default (no keyword). A public class is visible to all other classes, meaning another class may interact with the class, doing such things as creating an instance of the class and accessing its methods. Default (sometimes referred to as package access) classes are only visible to other classes within the same package.

Extending Classes

One of the most useful concepts of object oriented programming is that a class can extend another class. Imagine that you program a class that controls a robot arm to move up and down. You decide to call this class Arm. Let's examine a very simplified version of what this class might look like:

```
1. import lejos.nxt.*;
2. class Arm {
3.     public void armUp() {
4.         Motor.B.forward();
5.     }
6.      public void armDown() {
7.         Motor.B.backward();
8.     }
9. }
```

Now imagine you create an enhanced robot arm that still moves up and down, but also has an attached claw that opens and closes. It would be preferable to reuse the code in the Arm class instead of starting from scratch. It is also cumbersome to copy and paste the Arm code because if you ever change or improve the Arm code you would have to recopy it.

Java provides an elegant solution by extending the Arm class to inherit all the functionality of Arm, as well the ability to add new methods and data for the claw. In our example we will call this new class ClawArm. ClawArm extends Arm so Arm is the *superclass* and ClawArm is the *subclass* (Figure 3-2). All classes in Java automatically extend the Object class, as you can see in the diagram. If you follow the hierarchy of any Java class, the class at the top is always Object.

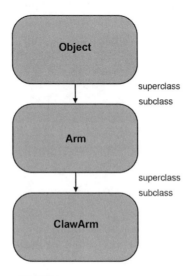

Figure 3-2 Superclass and Subclass

When a subclass extends another class, it inherits all of the functionality of the superclass. In Java code, the syntax for extending another class is as follows:

```
1. import lejos.nxt.*;
2. class ClawArm extends Arm {
3.     public void openClaw() {
4.         Motor.C.forward();
5.     }
6.
7.     public void closeClaw() {
8.         Motor.C.backward();
9.     }
10. }
```

The *extends* keyword is used above to indicate the superclass. Now anyone using the ClawArm object will be able to call the two methods of Arm, as well as the methods introduced in ClawArm:

```
ClawArm myRobot = new ClawArm();
myRobot.armDown();
myRobot.openClaw();
```

Abstract Classes

An abstract class (possibly) contains some functional methods, but also declares method names with no functional code. In a way, an abstract class is like a half-finished class; some methods are already provided, but others are just defined and must be filled in later. Since an abstract class is not complete it can not be instantiated. Let's examine an abstract class to see what they are about. We'll use the Arm class again:

```
 1.  import lejos.nxt.*;
 2.  abstract class Arm {
 3.      public abstract void spinArm();
 4.      public void armUp() {
 5.          Motor.B.forward();
 6.      }
 7.      public void armDown() {
 8.          Motor.B.backward();
 9.      }
10.  }
```

In line two, the Arm class is now declared to be Abstract. It contains the same two functional armUp() and armDown() methods as before, but it also defines a third abstract method called spinArm(). Notice this method has semicolons at the end of the definition, but no curly braces? If we tried to instantiate this class in code, the leJOS compiler will respond with "Arm is abstract; cannot be instantiated".

The purpose of an abstract class to be a higher level superclass that will be used by other subclasses. The subclasses all share the same code, making it more efficient and logical to program. So in this case, Arm is the general class, and the subclasses are more specific types of Arm (Figure 3-3).

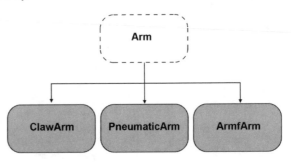

Figure 3-3 Hierarchy of Arm classes

Objects

Objects form the basis of object oriented programming. An object is an instance of a class that contains variables and methods. Put another way, they contain the data and functionality of a class.

A class is not a fully fledged object, but a programmer can use a class to create an object by using the *new* keyword to call a constructor on the class. The new object can also be assigned to a variable if the object needs to be referenced in later lines of code. The following lines of code show an object being initialized, and another assigning an object to a reference variable:

```
new ControlGUI(); // initialized but not assigned
String name = new String("Maximillion");
```

All Java classes are extended from the class Object. This class contains methods common to all Java objects, such as toString() and equals(). The toString() method provides a string representation of an object, and the equals() method compares two reference variables to see if they refer to the same object.

In Java, objects can be accessed by reference. This means two or more variables can refer to the same object, as the following pseudo-code demonstrates:

```
MyObject a = new MyObject();
MyObject b = a;
MyObject c = b;
```

As you can see above, all three reference variables (a, b, and c) are now referring to the same object. Imagine that a method is called using variable b, such as:

```
b.setValue(25);
```

This means the value will be changed for the object that a and c refer to as well, since they all refer to the same object. Now if you call getValue() on a, it will retrieve a value of 25:

```
int x = a.getValue(); // x = 25
```

Interfaces

Java does not allow a class to extend more than one class. Normally, with object oriented design, it is a good idea for a class to be responsible for only one type of behavior. If you try to make a single class perform all kinds of tasks it usually makes things unnecessarily complex.

But sometimes it is necessary to make a class with multiple behaviors. In these cases, interfaces are the solution. An interface can define methods and variables, but the methods do not contain any functional code. Since there is no functional code, an interface may not be instantiated (similar to an abstract class). The following is an example of an interface:

```
1. interface Steerable {
2.     public void turnLeft();
3.     public void turnRight();
4.     public void driveForward();
5.     public void applyBrakes();
6. }
```

As you can see there is no functional code here, only method definitions. So how does this help you as a programmer? Imagine a class that contains a method to steer any type of robot around a racetrack. We'll call this class RaceCar. Imagine that the RaceCar class has a method that steers a car around the racetrack by using the steering and driving methods contained by Steerable objects. We'll say the method definition looks like this:

```
public void drive(Steerable myCar)
```

By having your robot code implement Steerable, it can be used by the RaceCar class. It can also extend another class if need be (but this is not necessary). In the following example, we will also have the class extend the Thread class:

```
1.  import lejos.nxt.*;
2.  class MyCar extends Thread implements Steerable {
3.      public void turnLeft() {
4.          Motor.B.backward();
5.          pause(200);
6.          Motor.B.stop();
7.      }
8.
9.      public void turnRight() {
10.         Motor.B.forward();
11.         pause(200);
12.         Motor.B.stop();
13.     }
14.
15.     public void driveForward() {
16.         Motor.A.forward();
17.         Motor.C.forward();
18.     }
19.
20.     public void applyBrakes() {
21.         Motor.A.stop();
22.         Motor.C.stop();
23.     }
24.
25.     public void run() {
26.         // Sensor watching code here
27.     }
28.
29.     private void pause(int mSeconds) {
30.         try{
```

```
31.          Thread.sleep(mSeconds);
32.       }catch(InterruptedException e){}
33.    }
34. }
```

As you can see, there are now complete method definitions for turnLeft(), turnRight(), driveForward() and applyBrakes(). Now the RaceCar class will be able to use the MyCar class and guide it around the track using a few lines of code to get things going:

```
MyCar speedy = new MyCar();
RaceCar.drive(speedy);
```

RaceCar now has an instance of MyCar, called speedy. RaceCar knows it has a Steerable, so it can now call the proper methods to drive the car around the track.

Import and Package Statements

You may have noticed the following statement at the top of the MyCar class:

```
import lejos.nxt.*;
```

This is an import statement responsible for allowing the MyCar class access to all classes contained in the lejos.nxt package. Keep in mind that this statement does not cause all these classes to load into memory. Only the classes used in your code are uploaded to the NXT.

In Chapter Two we discussed some of the packages in the leJOS API, such as java.navigation and java.nxt.comm. These packages contain an assortment of classes that perform functions according to the theme of the package. The main theme of lejos.nxt is that all the classes in the package give you access to the NXT platform, such as MotorPort and SensorPort.

These classes can not be accessed by your code until the import statement is used, however. The advantage of the package system is that other classes can be hidden as well as sorted, but called out of hiding when you need to use them. You can also import just a single class from a package. For example, it we only want to use the Motor class we can use the following syntax:

```
import lejos.nxt.Motor;
```

Conversely, we could also access a class in a package without using an import statement by drilling down to it. The following line could be used to call a method directly from Motor:

lejos.nxt.Motor.B.forward();

NOTE: *The java.lang package is always imported automatically by Java. This is because the classes used in this package are central to the use of the Java language. It contains such important classes as Thread, System, String and Math which are used frequently.*

So how can you make your own packages? First, you must create a directory structure that matches the package name. This directory structure must start at a classpath directory. For example, you may want to make a package called *robot.flying*. If you have included c:\Java in your classpath, make the following directory:

```
c:\Java\robot\flying
```

Adding classes to your package is easy. Let's say you are making a class called AirShip and you want it to appear in the package robot. flying. Just include the package declaration at the top of your source file:

```
1. package robot.flying;
2. import lejos.nxt.*;
3. class AirShip {
4.    /// Rest of code...
5. }
```

The package statement must appear before any other code otherwise the compiler will produce an error. Once this class is created, in order to import it, the class file must appear in the directory c:\Java\robot\ flying. Usually I just store the source code file in the same directory as the package classes so that when it compiles it automatically dumps it in here, but you could move the class file here on your own.

NOTE: *The main purpose of packages is to organize code. On large projects with hundreds of classes, this can be very important. When creating your own leJOS programs you might not use very many classes. For this reason it might not be useful to create your own packages, but you will still need to know how to import classes from other packages in the leJOS API.*

Methods

A method gives a class its functionality. All methods must be contained within a class, and there is no such thing as a global method in Java. Methods have two important defining characteristics – return types and arguments. Let's examine a typical Java method definition:

```
public int readDistance() {
    return Sensor.S2.readValue();
}
```

The first line of this method declares a return type of the primitive int, so by definition the code within the curly braces must return an integer value. If a method declares it will return a value but the code does not return anything, the compiler will produce an error. Methods can just as easily return objects too. As you can see above, the parentheses are empty and contain no arguments. Let's examine a method that uses arguments:

```
1. public void setMotors(Motor left, Motor right) {
2.     leftDrive = left;
3.     rightDrive = right;
4. }
```

The above method has no return type, as indicated by using the keyword void. This method uses two arguments, left and right. These arguments can be objects (in this case, two Motor objects) or primitives.

NOTE: *Recursion is allowed in leJOS. This is an advanced programming technique of having a method call itself (often several hundred times) before a criterion is satisfied and the methods all return. Recursion is very memory intensive however, and is not recommended with the NXT.*

Constructor Methods

A constructor method is a special method used to initialize an object when it is created. The code within a constructor can contain absolutely anything, but generally it is limited to setting variables and preparing the object so it will work when a programmer starts calling methods. The following class shows a properly defined constructor method:

```
1. class Kangabot {
2.     int jumpCentimeters;
3.     public Kangabot(int jumpDistance) {
4.         jumpCentimeters = jumpDistance;
5.     }
6.
7.     public static void main(String [] args) {
8.         Kangabot roo = new Kangabot(5);
9.     }
10. }
```

The constructor method starts at line 3. As you can see, it has the same name as the class name, and it has no return type (not even void). In this case the constructor is simply used to initialize a programmer-defined variable jumpCentimeters. All objects use a constructor method, but if you don't specifically create one then the object will have a default no-arguments constructor implicitly defined by Java.

NOTE: *When you call the constructor method from another body of code, that body of code will stop its execution until the constructor returns. This means that if your constructor starts running methods that never end, your program will effectively freeze. This is one of the classic errors new programmers make.*

Static Methods

A static method is a method that can be called without creating an instance of a class. To contrast static methods with member methods, lets examine the following code.

```
MyRobot merl = new MyRobot();
merl.attack();
```

The code above demonstrates how a method is normally called from an object. With static methods, however, there is no need to create an instance first. If we changed the attack() method above to a static method, we could call it straight from the class as follows:

```
MyRobot.attack();
```

There are many Java classes, such as the java.lang.Math, that contain exclusively static methods. Since no data needs to be kept in an object for Math methods to work, there is no reason to go to the trouble of creating an instance of Math. So methods are called directly from the class instead, as follows:

```
int result = Math.sin(0.5);
```

It is easy to write a static method, but there are two rules that must be obeyed in order for it to work properly:

- A static method may not directly call an instance (non-static) method
- A static method may not directly use an instance (non-static) variable

Overriding Methods

As we saw, when one class extends another class the subclass inherits all the methods from the superclass. But sometimes it is beneficial to change one of the existing superclass methods in order to provide the subclass with altered or enhanced functionality. For example, when you create a two jointed robotic arm, you also need to program a class with various methods to control the arm. Imagine that the simplified API for the arm looks like this:

```
1. class RoboArm {
2.
3.     public void goToPoint(int x, int y, int z) {
4.         // Code to move hand to 3-D coordinate…
5.     }
6.
7.     // More methods...
8. }
```

Assume that your code is able to move the hand to any coordinate in 3-dimendional space. Now imagine that you add an extra joint around the wrist area of the robot arm. This means you need to alter the program code in order to accommodate the physical change.

The object-oriented way to do this is to extend the RoboArm class and then replace the goToPoint() method with new code. Replacing an existing method defined in a superclass is called *overriding* a method. There's not much to it, as the following example demonstrates:

```
1. class SuperArm extends RoboArm {
2.    public void goToPoint(int x, int y, int z) {
3.        // New code to move 3-jointed arm
4.    }
5. }
```

Primitive Data Types

leJOS allows code to declare all eight primitive types (although only 6 are fully implemented – see notes below). There are 4 integral numbers, 2 floating-point numbers, a character type and and a boolean type. Table 3-1 shows these types and their sizes.

Keyword	Minimum	Maximum	Default	Bits
byte	-2^8	$2^8 - 1$	0	8
short	-2^{16}	$2^{16} - 1$	0	16
int	-2^{32}	$2^{32} - 1$	0	32
long	-2^{32}	$2^{32} - 1$	0	64
float	varies	varies	0.0	32
double	varies	varies	0.0	64
boolean	none	none	false	2
char	'\U00000'	'\U65535'	'\U00000'	16

Table 3-1 Java Primitives

NOTE: *leJOS gives the illusion of allowing all eight Java primitives, but in fact it only fully supports six. The 64 bit numbers long and double are actually not completely supported, although you can still declare variables as long and double. For example, even though your code may look like it created a double value, the underlying primitive is a float. Likewise, any long literals are truncated to int.*

In some languages, such as Visual Basic, it is possible to declare a variable without specifying its type. Java, on the other hand, is a *strongly typed language*, which means all variables must be declared as belonging to a specific data type. The following class shows each of the primitive variables declared and initialized with a value:

```
1. class Primitives {
2.
3.     byte b = 127;
4.     short s = 32767;
5.     int i = 2147483647;
6.     long l = 2147483647;
7.
8.     float f = 100.123f;
9.     double d = 100.123;
10.
11.     char c = 'M';
12.
13.     boolean boo = true;
14. }
```

All of these variables are declared at the class level. Note in line 8 it is necessary to place an 'f' after the literal number. This is because literals are treated as double values by default, and it is illegal to assign a double value to a float variable.

NOTE: *Primitives are passed into methods (as arguments) by value. When a variable is passed in this manner it is just as if a copy of the variable was passed into the method. If the value is changed within the method it will not be reflected outside of the scope of the method.*

```
1. import lejos.nxt.*;
2.
3. class ByValue {
4.     public static void main(String [] args) {
5.         float height = 72f;
6.         float metricHeight = getCentimeters(height);
7.         LCD.showNumber((int)height);
8.     }
9.
10.     public static float getCentimeters(float inches) {
11.         inches = inches * 2.6f;
12.         return inches;
13.     }
14. }
```

The code above passes the value of variable height to the method getCentimeters. Inside the method, the value of the variable is multiplied by 2.6, but this does not mean the actual variable height is also multiplied. Since a copy of the variable is passed to the method, height is not affected.

Sometimes it is necessary to convert from one type of primitive to another. This can be done by casting, as when a ceramics artist casts plaster into a shape by using a mold. In Java, casting can convert a smaller primitive to a larger number type, or conversely, a larger value to a smaller primitive type. Let's examine how both of these look in code:

```
short x = 500;
int y = x; // implicit cast
int a = 500;
short b = (short)a; // explicit cast
```

Notice above that to convert a short number to an int requires no special syntax. Why? Because a smaller 16 bit primitive (short) will always fit within a larger 32 bit primitive (int). (If you have a 16 gallon jug and a 32 gallon jug, the contents of the 16 gallon jug will always fit in the 32 gallon jug no matter how full the 16 gallon jug is.) However, in lines 3 and 4, in order to convert the int to a short it must be explicitly stated that the conversion is occurring. If the number is too large to store in the smaller primitive then the number will be truncated.

It is also possible to cast floating-point numbers into integrals and vice versa. When converting a float to an int, the decimal places will be chopped off out of necessity. The following code demonstrates this:

```
float a = 555.555f;
int b = (int)a; // explicit cast
int x = 25;
float y = x; // implicit cast
```

Arrays

An array is a collection of objects or primitives that can be accessed using an index number. An array is an object in every sense of the word. It contains all the methods of the Object class such as equals() and toString(). It also has a variable called length that indicates how many elements it contains. The only difference is that it is initialized in a different manner from a regular object or primitive as the following examples demonstrate:

```
int [] x = new int[20];
x[5] = 2000; // changes the 6th value in the array
boolean [] boo = {true, false, false, false, true, false};
```

If you pass an array into a method as an argument and change one of the variables in the array, this will also be reflected outside of the method. So, like any object, an array is accessed by reference.

Naming Rules

As a programmer it is up to you to give your classes, methods, and variables unique names. There are several rules about naming these elements in code:

- The name must begin with a letter (upper or lower case), an underscore '_', or a dollar sign '$'

- The name may contain only characters, numbers, underscore or dollar sign.
- The name may not be a keyword (such as true, false, or null). Table 3-2 shows a complete list of keywords.

abstract	default	if	private	this
boolean	do	implements	protected	throw
break	double	import	public	throws
byte	else	instanceof	return	transient
case	extends	int	short	try
catch	final	interface	static	void
char	finally	long	strictfp	volatile
class	float	native	super	while
const	for	new	switch	
continue	goto	package	synchronized	

Table 3-2 Java Keywords

Operators

Operators lie at the very core of programming because they perform the actual calculations, which is what a computer is designed to do. Operators essentially perform low level math calculations involving bits.

Mathematical Operators

The basic mathematical operators usually function the same across all programming languages, so most people are quite familiar with them. They are:

- + addition
- - subtraction
- * multiplication
- / division
- % remainder (modulo division)

The most unfamiliar operation here for most new programmers is the modulo operator. It produces the remainder of integral number division; that is, the amount that was not able to divide evenly into a number and was thus left over. Modulo can be used for both integer numbers and floating point numbers. Examine the following code:

```
1. class MathTest {
2.     public static void main(String [] args) {
```

```
3.          int result1 = 15 % 2; // yields 1
4.          float result2 = 15 % 2.3f; // yields 1.2
5.      }
6. }
```

TIP: *If you have a hard time remembering whether to use the backslash \ or the forward slash / for division, remember the division symbol tilts in the same direction as the % symbol on the keyboard.*

NOTE: *There is no power operator in Java, such as the calculation 3^9. In order to perform a power calculation you must use the Math.pow() method, as follows:*

```
Math.pow(3, 9)
```

Comparison Operators

Comparison operators are used to compare two variables. All comparison operators return a boolean value to indicate if the comparison is true or not. Comparison operators are most often used in looping constructs, such as if-then and while (see below). The following comparison operators are used in Java:

```
==      // equals
>=      // greater than or equal to
<=      // less than or equal to
>       // greater than
<       // less than
!=      // not equal
```

Comparisons can be made as follows:

```
boolean a = 25 > 24; // true
boolean b = 25 == 24; // false
```

Boolean Operators

Some operators are specifically for numbers (greater than, less than), some are for boolean comparisons only (&& and ||), and some can be used for both numbers and boolean (!= and ==). The following can be used with boolean operands:

```
&&      AND
||      OR
^       Exclusive OR (XOR)
!=      Not equal
==      Equal
```

Let's examine these in some code:

```
1. boolean a = (25>24)||(12==13); // true
2. boolean b = (25>24)&&(12==13); // false
```

```
3. boolean c = a == b; // false
4. boolean d = a != b; // true
5. boolean e = a ^ b; // true
6. boolean f = b ^ c; // false
```

Most programmers are familiar with AND, OR, EQUAL and NOT EQUAL. Exclusive OR (XOR) is different from OR, however. Let's first compare this with OR. OR is true if one or the other, or both operands are true. XOR is only true if one or the other are true. If both are true, then XOR produces false.

Program Flow Control

A program that executed statements one after another in the same order each time it was run would be rather boring and predictable. It is the branching quality of a program that gives it power and flexibility. This quality is known as program flow control.

If Statements

If statements are very easy to use and very powerful. A typical if statement examines a boolean value and executes a block of code if the boolean value is true. The following example is typical example of an if-statement in leJOS code:

```
if(x == 10) {
   Sound.beep();
 }
```

Notice that no then keyword is used. All conditional code must appear in the curly braces. Alternatively, if there is only one statement to execute, the braces are not required:

```
if(x == 10) Sound.beep();
```

You can also use an else statement to execute a block of code should the boolean value equal false:

```
1. if(x == 10) {
2.    x = 0;
3.    Sound.beep();
4. } else {
5.    Sound.buzz();
6. }
```

Any code can be placed within the conditional code, including other if statements. This type of code construct is called a nested if-statement.

```
if(x == 10) {
   if(y == 5)
      Sound.beep();
 }
```

NOTE: *Switch statements are currently not supported by leJOS, but these can easily be replaced by if-else statements. For example:*

```
1. if (c == 'a') {
2.     ...
3. } else if (c == 'b') {
4.     ...
5. } else if (c == 'c') {
6.     ...
7. }
8.     ...
9. } else {
10.     // default behavior
11. }
```

For Loops

Conditional loops are used to repeat a code block a number of times until a condition is met. One of the most popular loop constructs is the for-loop. This loop repeats a block of code a predetermined number of times until a condition is satisfied. A for-loop consists of three main parts.

- Counter initialization
- Boolean condition
- Counter increment

These parts are stated in the following order:

```
for(counter initialization; boolean condition; counter
increment)
```

Let's examine a for-loop in some actual code:

```
1. import lejos.nxt.*;
2. class SoundLoop {
3.
4.     public static void main(String [] args) {
5.         for(int freq=500;freq<1000;freq += 50) {
6.             Sound.playTone(freq, 30);
7.         }
8.         try{
9.             Button.RUN.waitForPressAndRelease();
10.         } catch(InterruptedException e) {}
11.     }
12. }
```

The preceding example declares and initializes a variable called freq, checks if the variable satisfies the condition, executes the block of code, then increments the freq integer by 50 and rechecks the condition until the condition evaluates to false.

NOTE: *If the boolean condition is empty, it's assumed to be always true, so it will repeat in an endless loop as the following code demonstrates:*

```
for(;;) {}
```

While and do-while Loops

While loops are actually very similar to for loops, only they are not constructed specifically for incrementing a variable a set number of times. The while loop evaluates one boolean value, as follows:

```
while(boolean condition)
```

The following code uses a while loop to keep a robot moving forward until it gets to a dark area:

```
1.  import lejos.nxt.*;
2.
3.  class CockroachBot {
4.
5.      public static void main(String [] args) {
6.          LightSensor ls = new LightSensor(SensorPort.S2);
7.          Motor.B.forward();
8.          Motor.C.forward();
9.          while(ls.readValue() > 55) {
10.             // Keep moving forward
11.         }
12.         Motor.B.stop();
13.         Motor.C.stop();
14.     }
15. }
```

It is also possible to make the while loop execute the block of code at least once, then evaluate the condition. This is done using the do keyword, as follows:

```
do {
    // code body
} while(boolean condition);
```

Exception Handling

Exception handling is one of the unique features of Java that makes it popular with programmers. With exception handling, the error checking part of your code can be segregated from the rest of the functional code. This makes code neater and easier to understand. Java accomplishes this by enclosing method calls that may throw exceptions within a try-catch block. The following pseudo-code shows how this works:

```
1. try {
2.    // method that may throw exception
3. }
4. catch(Exception Type) {
5.    // code to deal with exception
6. }
```

When a method is prone to throwing an exception the try block must be called. The error is dealt with in the catch block. There is a third, optional part called a finally block. The finally block is executed once either the try or catch block has finished executing:

```
1. try {
2.    // method that may throw exception
3. }
4. catch(Exception Type) {
5.    // code to deal with exception
6. }
7. finally{ // Optional
8.    // code that is always executed no matter what
9. }
```

For the most part, under leJOS, the only time you'll really deal with exceptions is when using the Thread.sleep() method, as follows:

```
1. try {
2.    Thread.sleep(100);
3. } catch(InterruptedException e) {
4.    interrupted = true;
5. }
```

The java.lang Package

The previous section dealt with Java language fundamentals. Now we can have a look at the actual Java API included in leJOS. The first package we will examine is the java.lang package.

Math

The Math class is the place to go when you need complex mathematical functions. Some of these functions can be very useful in robotics where it is necessary to keep track of distances, angles, and coordinates. The leJOS Math class contains every method in the standard Java 2 Math class except methods whose worth is dubious to robotics, such as IEEEremainder() and rint().

NOTE: *All methods in the leJOS Math class use double values, but internally they are converted to float numbers to perform the math.*

java.lang.Math

- `public static final double E`

    The double value that is closer than any other to e, the base of the natural logarithms.

- `public static final double PI`

    The double value that is closer than any other to p, the ratio of the circumference of a circle to its diameter.

- `public static double sin(double a)`

    Returns the trigonometric sine of an angle.

    Parameters
    a: Angle value in radians.

NOTE: *All of the trigonometry functions return values in radians, as does the standard Java 2 package. This means that instead of getting a value of 180 degrees, the value will equal pi, or about 3.1416 (360 degrees equals 2pi in radians). If you prefer working in degrees you can use the Math.toDegrees() method for quick conversions.*

- `public static double cos(double a)`

    Returns the trigonometric cosine of an angle.

    Parameters
    a: Angle value in radians.

- `public static double tan(double a)`

    Returns the trigonometric tangent of an angle.

    Parameters
    a: Angle value in radians.

- `public static double asin(double a)`

    Returns the arc sine of an angle, in the range of -pi/2 through pi/2.

    Parameters
    a: Angle value in radians.

- `public static double acos(double a)`

    Returns the arc cosine of an angle, in the range of 0.0 through pi.

    Parameters
    a: Angle value in radians.

- `public static double atan(double a)`

  Returns the arc tangent of an angle, in the range of -pi/2 through pi/2.

  Parameters
  `a`: Angle value in radians.

- `public static double toRadians(double angdeg)`

  Converts an angle measured in degrees to the equivalent angle measured in radians.

  Parameters
  `angdeg`: Angle value in degrees.

- `public static double toDegrees(double angrad)`

  Converts an angle measured in radians to the equivalent angle measured in degrees.

  Parameters
  `angrad`: Angle value in radians.

- `public static double exp(double a)`

  Returns the exponential number e (i.e., 2.718...) raised to the power of a double value.

  Parameters
  `a`: Double value.

- `public static double log(double a)`

  Returns the natural logarithm (base e) of a double value.

  Parameters
  `a`: Double value.

- `public static double sqrt(double a)`

  Returns the correctly rounded positive square root of a double value.

  Parameters
  `a`: Positive double value.

- `public static double ceil(double a)`

  Returns the smallest (closest to negative infinity) double value that is not less than the argument and is equal to a mathematical integer.

  Parameters
  `a`: Double value.

- `public static double floor(double a)`

  Returns the largest (closest to positive infinity) double value that is not greater than the argument and is equal to a mathematical integer.

  Parameters
  a: Double value.

- `public static double atan2(double a, double b)`

  The regular atan() method accepts a value calculated by using y/x. The problem is, if either x or y is negative, the result of the fraction will also be negative but will not give a clue to which quadrant the angle is in. The atan2() method converts rectangular coordinates x, y (b, a) to polar (r, theta). This method computes the phase theta by computing an arc tangent of a/b in the range of -pi to pi.

  Parameters
  a: The y value in a coordinate system.
  b: The x value in a coordinate system.

- `public static double pow(double a, double b)`

  Returns of value of the first argument raised to the power of the second argument.

  Parameters
  a: The base number.
  b: The power to raise the base number to.

- `public static int round(float a)`
  `public static long round(double a)`

  Returns the closest int to the argument. The result is rounded to an integer by adding 1/2, taking the floor of the result, and casting the result to type int. In other words, the result is equal to the value of the expression:

  `(int)Math.floor(a + 0.5f)`

  Parameters
  a: Value to round.

- `public static double random()`

  Returns a double value with a positive sign, greater than or equal to 0.0 and less than 1.0. Returned values are chosen pseudo-randomly with (approximately) uniform distribution from that range. When this method is called, it uses a static instance of a pseudorandom-number generator, exactly as if by the expression:

  `new java.util.Random();`

This method is properly synchronized to allow correct use by more than one thread.

Parameters
a: Double value.

- ```
  public static int abs(int a)
  public static double abs(double a)
  ```

    Returns the absolute value of an int value. If the argument is not negative, the argument is returned. If the argument is negative, the negation of the argument is returned. Note that if the argument is equal to the value of Integer.MIN_VALUE, the most negative representable int value, the result is that same value, which is negative.

    Parameters
    a: An int or double value.

- ```
  public static int max(int a, int b)
  public static double max(double a, double b)
  ```

    Returns the greater of two int values. That is, the result is the argument closer to the value of Integer.MAX_VALUE. If the arguments have the same value, the result is that same value.

    Parameters
    a: First number.
    b: Second number.

- ```
  public static int min(int a, int b)
  public static double min(double a, double b)
  ```

    Returns the smaller of two float values. That is, the result is the value closer to negative infinity. If the arguments have the same value, the result is that same value. If either value is NaN, then the result is NaN. The floating point comparison operators consider negative zero to be strictly smaller than positive zero. If one argument is positive zero and the other is negative zero, the result is negative zero.

    Parameters
    a: First number.
    b: Second number.

Object

In Java, all objects are subclasses of the Object class, even if the code does not declare that the class extends Object. If you follow the hierarchy of any class all the way to the top you will find the Object class. There are a eight methods in the Object class, of which six are important to a casual programmer: toString() and equals().
  ```

java.lang.Object

- `public boolean equals(Object aObject)`

  The equals method compares two reference variables and tests whether they refer to the same object. To compare two objects, call the method from one of them, and use the other object as an argument:

  boolean match = firstObject.equals(secondObject);

  Parameters
  `aObject`: Object to compare with.

- `public String toString()`

  Normally this method returns a string representation of an object, but in leJOS it will just return an empty string, unless you override this method yourself. Once again, strings are not very important in robotics so this would only waste memory if it was implemented.

Runtime

The Runtime class allows you to check the memory in the NXT. The following two methods show how this works:

```
1.     public static void showFreeMemory() {
2.         Runtime rt = Runtime.getRuntime();
3.         int free = (int)rt.freeMemory();
4.         LCD.drawInt(free, 2, 0);
5.     }
6.
7.     public static void showTotalMemory() {
8.         Runtime rt = Runtime.getRuntime();
9.         int total = (int)rt.totalMemory();
10.        LCD.drawInt(total, 2, 2);
11.     }
```

java.lang.Runtime

- `public static Runtime getRuntime()`

  This method returns an instance of Runtime since the freeMemory() and totalMemory() methods are not static.

- `public long freeMemory()`

  This method returns the amount of free memory in the heap. The heap is the amount of memory the user program has access to (this will be explained further in Chapter 12).

- `public long totalMemory()`

  This method returns the total memory of the heap.

String

The String class simply contains an array of char primitives and methods to access those characters. String is given special status in Java in that it is not necessary to use the new keyword to create a String object (although the option to use it is open):

```
String island = "Tahiti";
```

It is also possible to create a string by joining two other strings together as follows:

```
String island = "Pit" + "cairn";
```

java.lang.String

- `public char[] toCharArray()`

    Returns an array of characters representing the string.

- `public String toString()`

    Returns itself.

- `public static String valueOf(Object aObj)`

    Returns the string representation of the Object argument.

    Parameters
    `aObj`: Object to obtain string representation from..

StringBuffer

In Java, a String object is immutable, meaning characters in the string cannot be added or removed once the string is initialized. In order to create a new set of characters a new String object must be created, which uses memory.

The StringBuffer, on the other hand, can be modified after creation, so it is more flexible and has the potential to save memory. Most of the methods in the leJOS StringBuffer class have to do with appending characters to a character array.

java.lang.StringBuffer

- `StringBuffer append(boolean aBoolean)`
  `StringBuffer append(char aChar)`
  `StringBuffer append(int aInt)`
  `StringBuffer append(long aLong)`
  `StringBuffer append(float aFloat)`
  `StringBuffer append(double aDouble)`
  `StringBuffer append(String aString)`
  `StringBuffer append(Object aObject)`

    Used to append a data type to the StringBuffer. If an object is used, the string representation of the object is used by calling toString().

    Parameters
    `a`: Data to append to the String.

- public char [] toCharArray()

    Returns an array of characters representing the StringBuffer.

System

The System class allows a programmer to interact with the operating system and retrieve information from it. In leJOS the only thing that can be retrieved is the system time:

java.lang.System

- public static long currentTimeMillis()

    Returns the number of milliseconds since the NXT was turned on.

Threads

Threads allow a program to execute several pieces of code simultaneously. They are very useful for robotics programming because each thread can be used to control a separate behavior. For example, you can use one thread per sensor that needs to be monitored. A single thread could be used to monitor the light in a room, while another could monitor a touch sensor.

   You can also use threads to control different parts of your robot. For example, I could create one thread to control a gun turret or robot arm, and another thread to control wheel movement. Threads are very easy to create in Java; you simply extend the abstract Thread class and place the main code for your thread in the run() method. In order to demonstrate this, let's make a program that does two tasks at once - counting to 1000 and playing random music:

```
1. import lejos.nxt.*;
2.
3. class BadMusic extends Thread {
4.
5.     public static void main(String [] args) {
6.         new BadMusic().start();
7.         new Counting().start();
8.     }
9.
10.     public void run() {
11.         while(true) {
12.             int freq = (int)(Math.random() * 1000);
13.             int delay = (int)(Math.random() * 40) + 10;
14.             Sound.playTone(freq, delay);
15.         }
16.     }
17. }
18.
19. class Counting extends Thread {
20.     public void run() {
21.         for(int i=0;i<1000;++i) {
```

```
22.              LCD.drawInt(i, 0, 2);
23.              LCD.refresh();
24.              try{Thread.sleep(1000);
25.              } catch(Exception e) {}
26.
27.          }
28.      }
29. }
```

The example above plays random, disjointed, Shatneresque "music" and is sure to be a favorite for many years to come. As you can see in this code, the program has two threads: BadMusic and Counting. BadMusic plays a random note for a random duration, one after another, in a never ending loop. The Counting class counts from zero to 1000, pausing for a second after each number.

Both of these methods are started in the main() method. Keep in mind that the main() method is its own thread, often called the primordial thread, so when you create a thread it will run concurrently with the main() code. Even though the main() method ends, the program will not terminate because the other two threads are still alive.

TIP: *Threads use a surprisingly large amount of memory so go easy on them and try to keep the number below eight.*

NOTE: *There is no Runnable interface in leJOS as there is in standard Java. This means all threads must extend the abstract Thread class and override the run() method.*

java.lang.Thread

- `public static Thread currentThread()`

    Returns an instance of the current thread, which will of course be the thread the line of code is running in.

- `public int getPriority()`

    Returns the priority of a Thread object.

- `public boolean isAlive()`

    Tests if this thread is alive.

- `public boolean isDaemon()`

    Checks whether this thread is a daemon thread (i.e. not a user created thread).

- `public boolean isInterrupted()`

    Checks whether this thread has been interrupted.

- `public abstract void run()`

  This method should be implemented with the main code for the thread.

- `public void setDaemon(boolean on)`

  Marks this thread as either a daemon thread or a user thread.

- `public void setPriority(int priority)`

  Sets the priority of the thread. Use the Thread constants to set the priority.

- `public static void sleep(long milliseconds)`

  Causes the thread to pause for a specific time.

- `public void start()`

  Begins the thread execution. (Remember not to call run() to start the method!)

- `public static void yield()`

  Causes the current thread to give way for another thread.

 NOTE: *java.lang.Class and java.lang.Clonable are not at all functional in leJOS. They are present merely because compilers require them to function properly.*

The leJOS NXJ API

Topics in this Chapter

- lejos.nxt
- lejos.nxt.comm
- javax.microedition.io
- java.io
- lejos.navigation
- lejos.subsumption
- javax.microedition.lcdui

Chapter 4

I magine a tiny robot that wanders around your house. In the past, such an activity was short lived because the robot invariably became stuck. It would tip over, run into a wall without the bumper activating, or the wheels would become stuck on some low-lying object.

Now imagine the following robot: it wanders around your house avoiding objects with the distance sensor. If the sensor misses an object, the robot can still tell if the wheels are stuck by monitoring decreases in rotation speed. If the robot tips over it uses a tilt sensor to identify the problem. It can even use speakers to emit a tiny voice that says, "I fell over on my side." It can then try to right itself. Such a robot could be left alone for hours, and when you return it would still be exploring your house. This is possible with leJOS NXJ. You just need to know where to find the methods in the API.

NOTE: *The leJOS developers periodically modify the API. Please refer to www.lejos.org for the latest API documentation (see Figure 4-1).*

Figure 4-1 Viewing the online leJOS API documentation

lejos.nxt

The classes and methods in the lejos.nxt package directly access the functions of the NXT brick.

Battery

The Battery class allows you to determine the voltage produced by the NXT batteries. Rechargeable batteries provide approximately 7.4 volts, while alkaline batteries produce 9 volts. When the voltage level begins to fall it means they are almost expended.

lejos.nxt.Battery

- `static float getVoltage()`

    Returns the battery voltage in volts.

- `static int getVoltageMilliVolt()`

    Returns the battery voltage in millivolts.

Button

The Button class contains static instances of the four buttons (much like the Motor class contains static instances of the motors). These four instances are ENTER, ESCAPE, LEFT, and RIGHT. Many times we just want the code to stop until a button is pressed, so we can use the waitForPressAndRelease() method:

```
// Stops code until ENTER pressed
try{
Button.ENTER.waitForPressAndRelease();
} catch(InterruptedException e) {}
```

You can also use a simple while-loop to stop your code while it waits for the user to press a button:

```
while(!Button.ENTER.isPressed()) {}
```

Java also offers event listeners. Java can initiate an action, or several actions, that are dependent on an event occurring (when a user presses a button, for example). There can be more than one listener waiting for an event to happen. When an event occurs, all the classes that are listening will be notified. The following example shows how to program an event listener:

```
1. import lejos.nxt.*;
2. class PlaySound implements ButtonListener {
3.     public void buttonPressed(Button b) {
4.         Sound.beepSequence();
5.     }
6.     public void buttonReleased(Button b) {}
7. }
```

This class implements the ButtonListener interface, which contains two method definitions: buttonPressed() and buttonReleased(). All interface methods must be defined in the class implementing the interface. When the button that is registered with this listener is pressed, the NXT will play a series of beeps. Now let's examine a class that registers this listener:

```
1. import lejos.nxt.*;
2. class ButtonTest {
3.     public static void main(String [] args) {
4.         Button.ENTER.addButtonListener(new PlaySound());
5.         while(true){} // Never ending loop
6.     }
7. }
```

As you can see in line four, the ENTER button has an instance of the PlaySound listener registered with it (up to four button listeners can also be registered for each button). The next line puts our main() method into a never ending loop, but it could just as easily have continued our program. We could also have opted to do this in one class by having ButtonTest implement the ButtonListener interface, and then add itself to the ENTER button.

- void addButtonListener(ButtonListener aListener)

    Adds a listener of button events.

- void callListeners()

    Call Button Listeners.

- boolean isPressed()

    Check if the button is pressed.

- void waitForPressAndRelease()

    Wait until the button is released.

ColorSensor

The ColorSensor class is used for accessing the HiTechnic color sensor (see Appendix A for information). It produces RGB color values, or it can identify colors from a palette.

lejos.nxt.ColorSensor

- int getColorNumber()

    Returns the color index detected by the sensor. The color index is shown in table 4-1.

0	Black	7	Orange
1	Violet	8	Red
2	Purple	9	Crimson
3	Blue	10	Magenta
4	Green	11–16	Pastels
5	Lime	17	White
6	Yellow		

Table 4-1 Index numbers of sensor color palette

- `int getBlue()`

    Returns the blue saturation of the color. (0 – 255)

- `int getGreen()`

    Returns the green saturation of the color. (0 – 255)

- `int getRed()`

    Returns the red saturation of the color. (0 – 255)

CompassSensor

The CompassSensor class works with either the HiTechnic or Mindsensors.com compasses (see Appendix A for information).

lejos.nxt.CompassSensor

- `public float getDegrees()`

    Returns the directional heading in degrees. (0 to 359.9) 0 is due North. On the Mindsensors circuit board a white arrow indicates compass direction. The black face on the HiTechnic sensor indicates the compass direction. The readings increase clockwise.

- `public float getDegreesCartesian()`

    This is an alternate method of retrieving compass direction, mainly used by the navigator classes. Compass readings increase clockwise, but Cartesian coordinate systems increase counter-clockwise. This method returns the Cartesian compass reading.

- `public void resetCartesianZero()`

    Changes the current direction the compass is facing into the zero angle. Affects only getCartesianDegrees()

- public void startCalibration()

    This method starts calibration for the compasses. When this is called, you must rotate the robot *very* slowly, taking at least 20 seconds per rotation. The Mindsensors compass requires at least two full rotations, while the HiTechnic requires one and a half to two rotations.

- public void stopCalibration()

    Ends calibration sequence.

NOTE: *Chapter 12 demonstrates how to calibrate a compass sensor.*

I2CSensor

I2CSensor is the abstract superclass of all I²C sensors, including UltrasonicSensor, CompassSensor, ColorSensor, TiltSensor and RCXLink. This means all the methods below can be called from these sensors. All sensors accept a SensorPort object in the constructor (e.g. SensorPort.S1). Unless you are making your own I²C sensors, the most useful methods are for reading sensor type, product ID and version numbers.

lejos.nxt.I2CSensor

- public int getData(int register, byte[] buf, int len)

    If you are making your own I²C class, use this to read data from the sensor.

    Parameters
    register: I2C register, e.g 0x41
    buf: Buffer to return data
    len: Length of the return data

- public int sendData(int register, byte[] buf, int len)

    Use this to send or change data on the sensor.

    Parameters
    register: I2C register, e.g 0x42
    buf: Buffer containing data to send
    len: Length of data to send

- public String getVersion()

    Returns the sensor version number.

- public String getProductID()

    Returns the sensor product identifier. Usually this is the maker of the sensor, such as "LEGO".

- public String getSensorType()

    Returns the sensor type. For example, "Sonar" for the ultrasonic sensor.

LCD

The LCD class offers some basic methods for drawing strings and numbers to the screen. For more advanced drawing functions, see the Graphics class in javax.microedition.lcdui.

- public static void drawString(String str, int x, int y)

    Displays a string on the LCD at a specified x, y coordinate. The coordinates start in the upper left corner. Each coordinate is one character in size. Nothing is displayed until refresh() is called.

- public static void drawInt(int i, int x, int y)

    Displays an integer on the LCD at specified x, y coordinate.

- public static void drawInt(int i, int places, int x, int y)

    This is a more advanced method for displaying integers. Characters from a previous call remain on the LCD. This method adds spaces to the front of the interger value, which will erase old characters. Displays an integer on the LCD at x, y with leading spaces to occupy at least the number of characters specified by the places parameter.

- public static void refresh()

    Updates the display. Nothing is displayed until this is called.

- public static void clear()

    Clears the display.

LightSensor

To use LightSensor you must create an instance of this class (see constructors below). The readValue() method returns a number between 0 and 100. Without calibration the values might not approach the low end or high end of this scale. Use calibrateLow() to set the zero level, and calibrateHigh() to set the 100 level. While using calibrateLow(), make sure it is completely in the dark. Point it at the brightest light in the room when using calibrateHigh().

NOTE: *All of the sensor constructors use SensorPort. However, if you look at the API it shows ADSensorPort or I2CPort. This architecture was adopted to differentiate between the two types of ports. SensorPort implements both of these types, which is why it is used in this chapter.*

lejos.nxt.LightSensor

- `public LightSensor(SensorPort port)`

  Creates a light sensor object attached to the specified port. The sensor will be set to floodlit mode, which turns on the red LED.

  Parameters
  port: e.g. SensorPort.S1

- `public LightSensor(SensorPort port, boolean floodlight)`

  Creates a light sensor object attached to the specified port, and sets floodlighting on or off.

  Parameters
  port: e.g. SensorPort.S1
  floodlight: true to set floodlight mode, false for ambient light.

- `public void setFloodlight(boolean floodlight)`

  Sets floodlighting on or off.

  Parameters
  floodlight: true to set floodlight mode, false for ambient light.

- `public int readValue()`

  Reads the current sensor value. Returns value as a percentage of difference between the low and high calibration values.

- `public int readNormalizedValue()`

  This method allows more accuracy than readValue(), since it returns a larger scale. Values can theoretically range from 0 to 1023, but typically they range from 145 (dark) to 890 (sunlight).

- `public void calibrateLow()`

  Call this method when the light sensor is reading the low value.

- `public void calibrateHigh()`

  Call this method when the light sensor is reading the high value.

Motor

Motor represents an NXT motor (not to be confused with the RCX motor). There are a lot more methods in the NXT motor class compared with the RCX, mostly because the built-in tachometer offers more functionality. Most of the additional methods are for indicating the current state of the motor (i.e. direction, speed).

This class contains three static instances of Motor: Motor.A, Motor.B and Motor.C. Speed is in degrees per second. The actual maximum speed of the motor depends on battery voltage and load. Let's examine the major methods of the Motor class.

TIP: *NXT robots have a hidden sensor to detect when they've hit an obstacle: the motors. That's because the Motor class has a getActualSpeed() method. If the motor is turning slower than it's supposed to, there's probably an obstacle hindering the robot and slowing down the motors. The project in Chapter 18 exploits this feature.*

lejos.nxt.Motor

- `public final boolean isMoving()`

    Returns true if the motor is in motion.

- `public final void stop()`

    Causes the motor to stop instantaneously. Once it is stopped, it resists any further motion.

- `public void flt()`

    Causes the motor to lose power and glide to a stop.

- `public void rotate(int angle)`

    Causes the motor to rotate the desired angle (in degrees).

- `public void rotate(int angle, boolean immediateReturn)`

    Causes the motor to rotate a desired angle. The method returns immediately and the motor will stop by itself when the angle is reached.

    Parameters
    `immediateReturn`: if true, method returns immediately.

- `public void rotateTo(int limitAngle)`

    Causes the motor to rotate to limitAngle. The tachometer should be within 2 degrees of the limit angle when the method returns.

- `public void rotateTo(int limitAngle, boolean immediateReturn)`

    This is the same as the method above, except with the option of returning immediately after the method call.

- `public void shutdown()`

    Disables the speed monitoring functions of this motor.

- `public void regulateSpeed(boolean isRegulated)`

    Turns speed regulation on/off.

- `public void smoothAcceleration(boolean isSmooth)`

    Enables smoother acceleration. Motor speed increases gently, and does not overshoot when regulateSpeed() is used.

- `public final void setSpeed(int speed)`

  Sets motor speed, in degrees per second. Up to 900 is possible with fully charged batteries.

- `public void setPower(int power)`

  Sets motor power. If speed regulation is enabled, the results of changing this are unpredictable.

  Parameters
  power: power setting: 0 - 100

- `public final int getSpeed()`

  Returns the speed this motor is set to, in degrees per second. Does not return the actual measured speed— see getActualSpeed().

- `public int getLimitAngle()`

  Returns the angle (in degrees) that a Motor is rotating to.

- `public final boolean isRotating()`

  Returns true when motor is rotating toward a specified angle.

- `public int getActualSpeed()`

  Returns actual speed, in degrees per second, calculated every 100 ms. A negative value means the motor is rotating backward.

- `public int getTachoCount()`

  Returns the tachometer count in degrees.

- `public void resetTachoCount()`

  Resets the tachometer count to zero.

- `public int getPower()`

  Returns the current power setting (0-100).

- `public void forward()`

  Causes the motor to rotate forward.

- `public boolean isForward()`

  Returns true if the motor is rotating forward.

- `public void backward()`

  Causes the motor to rotate backwards.

- `public boolean isBackward()`

  Returns true if the motor is rotating backward.

- `public void reverseDirection()`

  Reverses direction of the motor. It only has an effect if the motor is moving.

- `public boolean isFloating()`

    Returns true if the motor is in float mode.

- `public boolean isStopped()`

    Returns true if the motor is stopped.

MotorPort

The MotorPort class was added to the leJOS NXJ to handle backward compatibility with RCX motors, and to allow for multiplexer units which extend the number of motor ports. There are three static instances of MotorPort: A, B, and C. You can use one of these to construct motors, such as RCXMotor (see below).

`RCXMotor m = new RCXMotor(MotorPort.A);`

You are unlikely to call any of these methods directly with regular robotics projects, but if you are curious you can browse the leJOS NXJ API documentation.

RCXLightSensor

RCXLightSensor is the legacy version of the light sensor contained in the RIS kit. The regular LightSensor class does not work properly with the older RCX light sensor.

lejos.nxt.RCXLightSensor

- `public RCXLightSensor(SensorPort port)`

    Creates an RCX light sensor object attached to the specified port. The sensor will be activated, i.e. the LED will be turned on.

    Parameters
    port: port, e.g. SensorPort.S1

- `public void activate()`

    Activates an RCX light sensor. You should see the LED go on when you call this method.

- `public void passivate()`

    Passivates an RCX light sensor (turns off LED).

- `public int readValue()`

    Reads the current sensor value. Returns a percent value.

RCXLink

RCXLink is a class that allows the NXT brick to control an RCX brick via the IR port. There are RCX linking devices by HiTechnic and Mindsensors.com (see Appendix A).

The RCXLink class has methods for controlling the basic functions of the RCX brick. It contains three static instances of RCXMotor: A, B and C. They are accessed as follows:

```
RCXLink rcx = new RCXLink(SensorPort.S1);
rcx.A.forward();
```

lejos.nxt.RCXLink

- `public RCXLink(SensorPort s)`

    Initializes the RCX link.

- `public void runProgram(int programNumber)`

    Runs a program on the RCX.

    Parameters
    programNumber (1-5)

- `public void stopAllPrograms()`

    Stops the currently executing program on the RCX.

- `public void powerOff()`

    Sends command to turn off the RCX brick.

- `public void beep()`

    Makes the RCX chirp.

`public void setLongRange(boolean longrange)`

    Sets the range of the IR light on the RCXLink. Long range uses 25mA, short range 15mA.

    Parameters
    longrange (true = long range, false = short range)

- `public void setHighSpeed(boolean highspeed)`

    Sets the communications speed of the RCXLink. The default speed is 2400 baud.

    Parameters
    highspeed (true = 4800 baud, false = 2400 baud)

- `public void sendMacro(byte [] macroCommands)`

    Transmits a series of macro commands to the RCX. Up to 175 macro commands may be sent to the RCX.

- `public void runMacro(byte macro)`

    Run the ROM/EEPROM macro at address 0xXX

RCXMotor

RCXMotor is a class for legacy RCX motors. As you can see below, many of the methods are identical to Motor. To obtain an instance of this class, use a MotorPort in the constructor.

lejos.nxt.RCXMotor

- `public RCXMotor(MotorPort port)`
  Constructor for RCXMotor.

- `public void setPower(int power)`
  Sets power to the RCX motor.

  Parameters
  `power`: power setting (0 – 100)

- `public int getPower()`
  Returns the current power setting (0 – 100).

- `public void forward()`
  Causes motor to rotate forward.

- `public boolean isForward()`
  Return true if motor is rotating forward.

- `public void backward()`
  Causes motor to rotate backwards.

- `public boolean isBackward()`
  Return true if motor is rotating backward.

- `public void reverseDirection()`
  Reverses direction of the motor. It only has effect if the motor is moving.

- `public boolean isMoving()`
  Returns true if the motor is in motion.

- `public void flt()`
  Causes motor to float. The motor will lose all power, but this is not the same as stopping.

- `public boolean isFloating()`
  Returns true if the motor is in float mode.

- `public void stop()`
  Causes motor to stop instantaneously. The motor will resist any further motion.

- `public boolean isStopped()`

  Return true if motor is stopped.

- `public int getMode()`

  Returns the current mode of the motor.

Mode	Value
Forward	1
Backward	2
Stopped	3
Floating	4

Table 4-1 RCX modes

RCXMotorMultiplexer

The RCXMotorMultiplexer class allows access to the Mindsensors.com multiplexer for RCX motors. It is similar to RCXLink in that it contains instances of RCXMotor (four in this case). You can access these motors as follows:

```
RCXMotorMultiplexer expansion = new RCXMotorMultiplexer(
    SensorPort.S1);
expansion.A.forward();
```

The methods to control the each RCX motor are the same as those found in RCXMotor.

SensorPort

SensorPort is similar to MotorPort. There are four static SensorPort objects in this class. Generally these are used in the constructors for sensors. This class contains many methods for accessing data directly from the port, either using I²C protocol or analog sensors. You are unlikely to use many of them yourself when programming robots, but if you are interested in developing your own sensor classes you might want to examine the API more thoroughly. The one method you might use is to access the event model, as shown below.

lejos.nxt.SensorPort

- `addSensorPortListener(SensorPortListener aListener)`

  Adds a port listener to this port. aListener must implement the stateChanged() method in the interface SensorPortListener. You can add at most 8 listeners to a sensor port.

Sound

The Sound class is responsible for playing sounds through the NXT speaker. The playTone() method is the most versatile. There are also convenience methods for playing basic sounds.

lejos.nxt.Sound

- `public static void playTone(int aFrequency, int aDuration)`

  Plays a tone, given its frequency and duration. Frequency is audible from about 31 to 2100 Hertz. The duration argument is in hundredths of a second (centiseconds, not milliseconds) and is truncated at 256, so the maximum duration of a tone is 2.56 seconds.

  Parameters
  aFrequency: The frequency of the tone in Hertz (Hz).
  aDuration: The duration of the tone, in centiseconds. Value is truncated at 256 centiseconds.

- `public static void beep()`

  Beeps once

- `public static void twoBeeps()`

  Beeps twice

- `public static void beepSequence()`

  Downward tones

- `public static void buzz()`

  Low buzz

SoundSensor

The SoundSensor class allows access to the LEGO sound sensor. It has two modes of operation, dB and dBA mode. dB mode measures straight decibels, while dBA measures sound intensity weighted for the range of human hearing.

lejos.nxt.SoundSensor

- `public SoundSensor(SensorPort port)`

  Creates a sound sensor object attached to the specified port. The sensor will be set to DB mode.

  Parameters
  port: e.g. SensorPort.S1

- `public SoundSensor(SensorPort port, boolean dba)`

  Creates a sound sensor object attached to the specified port, and sets DB or DBA mode.

Parameters

port: e.g. SensorPort.S1

dba: true to set DBA mode, false for DB mode.

- `public void setDBA(boolean dba)`

  Set DB or DBA mode.

  Parameters

  dba: true to set DBA mode, false for DB mode.

- `public int readValue()`

  Read the current sensor value, as a percentage.

TiltSensor

The TiltSensor class allows access to the Mindsensors.com and HiTechnic tilt sensors (see Appendix A for information).

lejos.nxt.TiltSensor

- `public TiltSensor(SensorPort port)`

  `public int getXTilt()`

  Returns X tilt value.

- `public int getYTilt()`

  Returns Y tilt value.

- `public int getZTilt()`

  Returns Z tilt value.

- `public int getXAccel()`

  Returns positive or negative acceleration along the X axis as mg (mass x gravity). The variable g represents acceleration due to gravity (9.81 m/s^2). If this accelerometer was falling on earth and oriented along the X axis, it would return 9810 mg.

- `public int getYAccel()`

  Returns acceleration along the Y axis in mg.

- `public int getZAccel()`

  Returns acceleration along the Z axis in mg.

TouchSensor

Allows access to the basic NXT or RCX touch sensor.

lejos.nxt.TouchSensor

- `TouchSensor(SensorPort port)`

  Create a touch sensor object attached to the specified port.

- `boolean isPressed()`

    Check if the sensor is pressed.

UltrasonicSensor

The UltrasonicSensor class is used to obtain distance readings from the LEGO ultrasonic sensor.

TRY IT: *While the sensor is in use, place it next to your ear. You can hear rapid clicks continuously from the sensor, which sound almost like bat pings.*

- `public UltrasonicSensor(SensorPort port)`

    Creates an instance of UltrasonicSensor.

- `public int getDistance()`

    Returns the distance to an object. If no object is in range it returns 255.

lejos.nxt.comm

The leJOS communications API is covered in depth in Chapter 16 on networking. Briefly, it contains the following classes.

Bluetooth

The main class for establishing a Bluetooth connection. It is responsible for sending and receiving bytes.

USB

The main class for establishing a USB connection. It is responsible for sending and receiving bytes.

BTConnection

An object for obtaining input and output streams from Bluetooth. This class is a subclass of javax.microedition.io.StreamConnection.

USBConnection

An object for obtaining input and output streams from USB. This class is a subclass of javax.microedition.io.StreamConnection.

LCP

This class is the leJOS implementation of the LEGO Communications Protocol, which is used for communications between the PC and LEGO firmware. We've adopted this protocol in leJOS for controlling the NXT brick.

javax.microedition.io

This package contains a StreamConnection class (mentioned above) that allocates streams. Chapter 16 examines this package in more detail.

java.io

The java.io package contains the various streams used in Java communications. These include the basic InputStream and OutputStream classes, which can send only bytes. To send larger data types, such as integer and double values, use the DataInputStream and DataOutputStream classes. Chapter 16 examines this package in more detail.

lejos.navigation

The navigation API has changed since the previous version of leJOS for the RCX. Now there are Pilot classes, which control steering functions for robots using two synchronized motors. These classes do not keep track of position. The package also contains Navigators, which keep track of position. Chapter 12 explores the navigation API in detail.

lejos.subsumption

The subsumption package contains classes for implementing Rodney Brook's behavior control. There are actually two versions of this architecture in this package. The basic version uses a Behavior interface and an Arbitrator class. The second is more advanced and more extensible than the first, and it uses Activity and ActivityBase classes. Chapter 18 explores the basic subsumption architecture.

javax.microedition.lcdui

As part of Java Micro Edition, SUN Microsystems designed a package specifically for small devices. This package is targeted for devices that lack large color displays, lots of memory, hard drives, networking, and mouse/keyboard control.

There are several flavors available (called profiles by SUN), depending on the hardware capabilities of the device. One profile is specifically for mobile phones and other mobile devices. SUN calls this profile the Mobile Information Device Profile (MIDP). The MIDP contains a simple API tailored for systems with small LCD screens, very little memory, and limited input. The NXT hardware neatly fit into this profile, so the leJOS developers decided to adapt parts of this API for the NXT, including the Graphics class (below).

Graphics

The graphics class is used to draw a variety of primitive shapes to the LCD screen, such as circles, ovals, arcs, lines, and rectangles. Unlike the official Graphics class, this one draws directly to the LCD screen, rather than to a Panel or other component. You must first create an instance of Graphics before using the class:

```
1. Graphics g = new Graphics();
2. g.drawLine(5,5,60,60);
3. g.drawRect(62, 10, 25, 35);
4. g.refresh();
```

javax.microedition.lcdui.Graphics

- `public void setPixel(int color, int x, int y)`

    The color parameter is obtained from Graphics, and represents either black or white. Use Graphics.BLACK or Graphics.WHITE.

- `public void drawLine(int x0, int y0, int x1, int y1)`

    Draws a line between the points (x1, y1) and (x2, y2).

- `public void drawRect(int x, int y, int width, int height)`

    Draws a rectangle. The left and right edges of the rectangle are at x and x + width. The top and bottom edges are at y and y + height (see Figure 4-2).

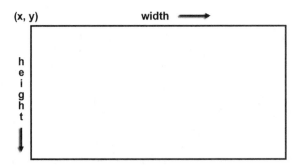

Figure 4-2 Drawing a rectangle

- `public void fillRect(int x, int y, int width, int height)`

    Draws a rectangle, as above, but fills it solid.

- `public void drawArc(int x, int y, int width, int height, int startAngle, int arcAngle)`

    The drawArc() method is used for drawing circles, ovals, and of course arcs. It draws the outline of a circular or elliptical arc covering the specified rectangle (see Figure 4-3). The resulting arc begins at startAngle and extends for arcAngle degrees. Angles are interpreted such that 0 degrees is at the 3 o'clock position. A positive value indicates a counter-clockwise rotation while a negative value indicates a clockwise rotation.

drawArc(10, 10, 40, 30, 5, 90)

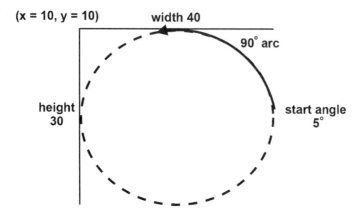

Figure 4-3 Drawing an arc

- public void fillArc(int x,int y,int width,int height,int startAngle,int arcAngle)

    The fillArc() method is used for drawing solid circles, ovals, and arcs. It fills in the area from the center of the arc to the outer line. It's useful for drawing Pac-Man.

- public void drawRoundRect(int x,int y,int width, int height, int arcWidth, int arcHeight)

    Draws a round-cornered rectangle (see Figure 4-4). The left and right edges of the rectangle are at x and x + width, respectively. The top and bottom edges of the rectangle are at y and y + height. The arcWidth and arcHeight parameters determine the size of the rounded corners.

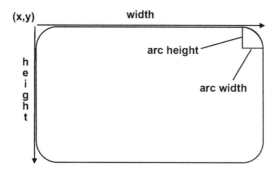

Figure 4-4 Drawing a round rectangle

- `public void refresh()`

    Updates the LCD display. Nothing will appear on the LCD screen until refresh() is called.

- `public void clear()`

    Clears the LCD display buffer. Must call refresh() before this method shows up.

This takes care of the boring stuff. Now you can move on to more exciting things, like programming robots. Hopefully this overview will allow you to create the robot you have been dreaming of.

LEGO Parts

Topics in this Chapter

- Beams
- Liftarms
- Pins
- Axles
- Axle Accessories
- Tires and Wheels
- Gears
- Cables
- Other Parts

Chapter 5

Y ou become an artist when you build a new creation. Like a sculptor who creates art with clay, you can use LEGO parts to sculpt your own unique art. LEGO gives you the ability to design art with function. The more useful the function, the more valuable your art becomes. This chapter will review the basic components of the NXT kit, and give you tips on how to get the best result with these parts.

NXT Parts Overview

NXT parts are incredibly versatile and, when combined with an active imagination, can be used to build many different machines. This book alone describes plans for over a dozen diverse projects. Yet, despite how dissimilar these machines may be, the basic parts used to create them are the same.

Most parts in the NXT kit are LEGO TECHNIC™ parts. The product line was first introduced in 1977 as the *Expert Builder* series and was renamed as the TECHNIC series in 1986. These parts differ from the standard LEGO bricks in that they can be used to build complex moving machines.

All the parts in your NXT kit are of high quality. The accuracy of LEGO's injection molding process is seemingly flawless, with no defects or vestigial plastic. On top of that, the quality of the plastic is impressive. Each part appears to be almost indestructible, and won't easily scratch, dent, bend or break. I've been especially hard on the gears and, surprisingly, they never shatter or wear down.

While users will be impressed by the improvements of the NXT brick, they might not be so impressed with the parts selection compared to the RIS kit. There are now only 577 parts. Of these, only 81 are unique, whereas the RIS kit has 141 unique parts.

WEBSITE: *All LEGO parts and sets have a part number. You can look up the NXT kit by its number (8527) on the* Peeron® *web site and see all the numbered parts. You can also look up a particular part to determine which LEGO set it belongs to, as well as sources to purchase individual parts. Visit* **www.peeron.com.**

Most of the parts in the LEGO kit were chosen based on the four demonstration models in *Robo Center*. *Alpha-Rex*, the showcase robot on the cover of the box, had the most influence on the parts selection. If you build *Alpha-Rex* you will notice it uses the exact number of many parts in the kit.

The biggest challenge for first time LEGO builders is to figure out what all the odd-shaped parts are capable of building. Let's look through the parts contained in your kit and see what they can do.

NOTE: *Part names are taken from the LDraw part reference, which is more consistent with naming than Peeron. The number following some parts refers to the length in LEGO units. Sometimes the name itself has a number followed with the letter "L", which also refers to LEGO units. The number in parenthesis refers to the number of parts in the NXT kit. The number after the brackets is the part number.*

E: Beam 11   (7 white)   32525

name   length   qty.   color   part #

Beams

A: Beam 3	(16 dark grey)	32523
B: Beam 5	(5 dark grey)	32316
C: Beam 7	(6 white)	32524
D: Beam 9	(7 white)	120
E: Beam 11	(7 white)	32525
F: Beam 13	(4 white)	41239
G: Beam 15	(11 white)	32278

Beams (sometimes called girders) form the chassis of your robotic
creations. They have no studs, only axle holes. When beams are bent
they are called liftarms (see section below). All lengths are an odd
number of LEGO units, so most of your creations will have holes along
the center axis.

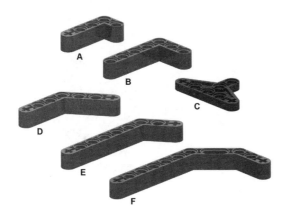

Liftarms

A:	Liftarm 5 Bent 90	(10 dark grey)	32140
B:	Liftarm 7 Bent 90	(8 dark grey)	32526
C:	Liftarm Triangle 5 x 3 x 0.5	(4 black)	2905
D:	Liftarm 1x7 Bent 53.5	(16 dark grey)	32348
E:	Liftarm 9 Bent 53.5	(2 dark grey)	32271
F:	Liftarm 11.5 Bent 45 Double	(6 dark grey)	32009

Liftarms are useful for altering the angle of your construction and
attaching motors to the NXT brick. The NXT kit contains no studs
(other than on a few token bricks), which means you will need to find
other ways to stack beams on top of other beams. Liftarms are ideal
for constructing boxes with beams.

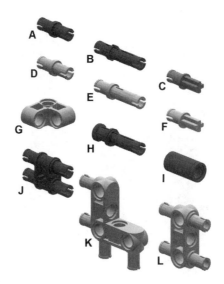

Pins

A:	Pin with Friction	(80 black)	4459
B:	Pin Long with Friction	(34 black)	6558
C:	Axle Pin with Friction	(42 blue)	43093
D:	Pin	(2 grey)	3673
E:	Pin Long	(13 grey)	32556
F:	Axle Pin	(4 yellow)	3749
G:	Pin Joiner Perpendicular Bent	(1 grey)	44809
H:	Pin Long with Stop Bush	(8 black)	32054
I:	Pin Joiner Round	(4 black)	75535
J:	Pin 3L Double	(3 black)	32136
K:	Beam 5 Bent 90 with 4 Pins	(8 grey)	55615
L:	Axle Joiner Perpendicular with 4 Pins	(13 grey)	48989

TECHNIC pins are the glue that holds your robot together. These pins are inserted in the holes of beams to connect to other beams. There are more black friction pins in the NXT kit than any other part, which is a clear indication that you will use a lot of these in everything you build. Blue axle pins with friction are useful when you need to join a beam to an X-shaped axle hole.

The grey non-friction pins are used when you need a joint between two beams that moves freely. The RIS 2.0 kit contained 24 of them, but now there are only two. If you need more than two non-friction pins, you will have to rely on the black friction pins(which don't move as freely), or the long grey non-friction pins.

There are also yellow non-friction axle pins, which are useful for attaching idler gears or wheels and liftarms to beams. Again, there are very few of these pins. The RIS 2.0 kit contained 16, but now there are only four to work with.

Some pins are attached to beams and liftarms which help connect beams together at different angles and spacing. You will find these very useful in most robots.

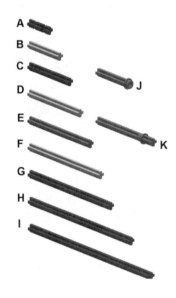

Axles

A: Axle 2 Notched	(22 black)	32062
B: Axle 3	(16 grey)	4519
C: Axle 4	(4 black)	3705
D: Axle 5	(7 grey)	32073
E: Axle 6	(4 black)	3706
F: Axle 7	(4 grey)	44294
G: Axle 8	(2 black)	3707
H: Axle 10	(4 black)	3737
I: Axle 12	(2 black)	3708
J: Axle 3 with Stud	(4 dark grey)	6587
K: Axle 5.5 with Stop	(2 dark grey)	32209

Axles are used for rotating parts, but occasionally can be used to connect a series of beams together. They can also be used to create legs, arms, or other specialized items. The color of the axle has no bearing on the function; they all exert the same friction.

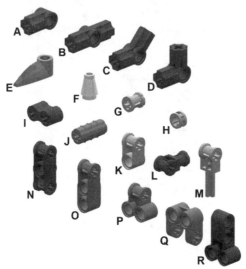

Axle Accessories

A: Angle Connector #1	(4 black)	32013
B: Angle Connector #2	2 black)	32034
C: 135 Degree Angle Connector #4	(2 black)	32192
D: 90 Degree Angle Connector #6	(16 black)	32014
E: Bionicle Tooth with Axle Hole	(8 orange)	X346
F: Cone 1 x 1	(3 white)	4589
G: Bush	(16 grey)	3713
H: Bush smooth	(6 grey)	32123
I: Axle Joiner Double Flexible	(8 black)	45590
J: Axle Joiner	(2 dark grey)	6538A
K: Axle Joiner Perpendicular	(8 grey)	6536
L: Connector with Axle Hole	(4 black)	32039
M: Pole Reverser Handle	(1 grey)	6553
N: Axle Joiner Perpendicular 3L	(4 black)	32184
O: Axle Joiner Perpendicular with 2 Holes	(6 dark grey)	42003
P: Axle Joiner Perpendicular Double	(6 dark grey)	32291
Q: Axle Joiner Perpendicular Double Split	(6 dark grey)	41678
R: Pin Joiner Dual Perpendicular	(4 black)	32557

The LEGO NXT kit has a variety of accessories that can couple with axles. Angle connectors are used to support axles or to alter the angle of an axle. Each connector has an identity number that is used to identify the angle. Another part that can be used with axles is the rubbery, axle joiner. They are useful for the ends of legs because they can grip slippery surfaces.

Bushes are used to secure axles in place. Although the full bush has more surface area, the half-bush exerts about the same friction on an axle because the axle hole is smaller. If you need to make your robot extra secure in places, try using two half-bushes in place of a full bush. The orange Bionicle tooth is mostly decorative, but it can act as a bush if you run out of standard bushes.

Often you need a longer axle than provided in the NXT kit. Axle joiners allow you to connect two axles together. It might appear that the kit only contains two axle joiners to extend axle lengths. However, the angle connector #2 works almost as well as an axle joiner. If you are really pressed, there are other ways to extend axles (see figure 5-1).

Figure 5-1 Extending axles

My favorite parts are the double-perpendicular axle joiners. Together, these parts allow you to attach beams in a number of ways. There are two different double-perpendicular axle joiners that are complimentary to one another and fit together with an axle (see Figure 5-2: A and B). You can also fit a double-split perpendicular axle joiner with a dual-perpendicular pin joiner to create a door hinge. You can use a variety of other parts with the double-split perpendicular axle joiner to create other parts (see Figure 5-2: C through G).

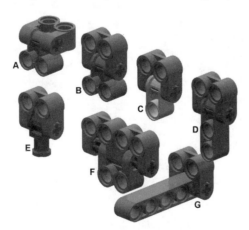

Figure 5-2 Double Perpendicular Axle Connectors

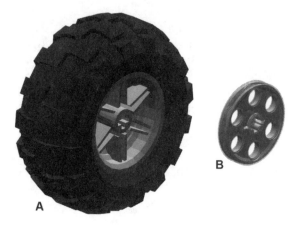

Tires and Wheels

A: Tire 56 x 26 Balloon   (4 black)   55976

     Wheel 43.2 x 22   (4 grey)   54087

B: Wedge Belt Wheel   (2 grey)   4185

There is little choice in the NXT kit when it comes to tires. The RIS Kit contained seven different wheel sizes, but there is only one size in the NXT kit. If you need a castor wheel, the wedge belt wheel can be used, as can double bevel gears.

The NXT kit does not contain tank treads. This is a major disappointment, especially if you like *Number-5* from the movie *Short Circuit* (1986, TriStar Pictures).

TIP: *If you want tank treads but don't have an RIS kit, try ordering the expansion kit 9648 mentioned in Chapter 1. The wheel included in the NXT has ribs on the outside of the hub that mesh with the tank treads included in kit 9648. If you want larger treads, the Snowmobile (8272) contains larger tank treads made from individual tread links. The kit also includes two shock absorbers.*

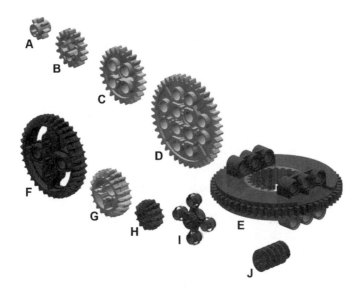

Gears

A:	Gear 8 Tooth	(6 grey)	3647
B:	Gear 16 Tooth	(2 grey)	4019
C:	Gear 24 Tooth	(7 grey)	3648
D:	Gear 40 Tooth	(1 grey)	3649
E:	TECHNIC Turntable Top	(1 dark grey)	48451
	TECHNIC Turntable Base	(1 black)	48452
F:	Gear 36 Tooth Double Bevel	(1 black)	X344
G:	Gear 20 Tooth Double Bevel	(2 grey)	32269
H:	Gear 12 Tooth Double Bevel	(4 black)	32270
I:	Knob Wheel	(4 black)	32072
J:	Worm Screw	(2 black)	4716

Torque is force in a rotational motion. A very powerful motor can produce a lot of torque, and likewise a weak motor produces low torque. Gears are used to transfer torque from one axle to another, often increasing or decreasing torque in the process. When two or more gears decrease speed, torque increases. When gears increase speed, torque decreases (torque and speed have an inverse relationship).

The square-tooth flat gears normally interlock with the small 8-tooth gear, so all larger gears (including the turntable) are divisible by 8 for easier gear reduction ratios (see Table 5-1). This is just one example of LEGO's brilliance with details.

Gear reduction occurs when a smaller gear connects to a larger gear. For example an 8-tooth gear on one axle will rotate three times for every one time a 24-tooth gear rotates. This is a 3:1 gear reduction.

Gear	Ratio with 8-tooth gear
8-tooth	1
16-tooth	2
24-tooth	3
40-tooth	5

Table 5-1 Determining gear ratios

Gear reduction can be used to make your tachometer more accurate. If it takes 60 counts to rotate the arm into a position, and your count is off by 2, this will reduce arm movement accuracy by 3% If, however, it takes 600 counts to move the arm, and the count is off by 2, the loss in accuracy is only 0.3%.

LEGO provides special gears to change the axis of rotation by 90 degrees. There are bevel gears, which appear fatter and rounder than regular gears, and there are knob wheels, which look like tire irons. The bevel gears mesh with one another, but they must be spaced properly, sometimes using half-bushes (see Figure 5-3).

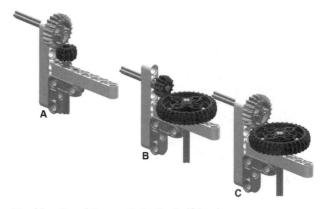

Figure 5-3 Meshing Bevel Gears. Note the half-bushes

Knob wheels perform the same function as bevel gears by changing the axis of rotation. Knob wheels are superior to the bevel gears, which tend to slip when under large forces. If you are creating a gear transfer that has substantial force on it, such as an arm or rough-terrain vehicle, use knob wheels.

TIP: There are only 4 valuable knob-wheels in the kit, but you can stretch this number out to 6 by placing towball pins in the holes of the motor (see figure 5-4).

Figure 5-4 Using towball pins to replace knob wheels

Worm screws are another part that can be used to change the axis of rotation. Worm screws can also be used for gear reduction. In fact, they are the most effective gear reducers; one complete rotation of a worm screw moves any gear by only one tooth. Rotation can only travel from the worm screw to regular gears, but regular gears cannot turn the worm screw (it locks if you try).

The turntable is generally used for rotating larger structures, such as a crane. The inner gear (grey) has 24 teeth and can mesh with an 8-tooth gear. The outer gear (black) has 56 teeth and can mesh with larger gears.

TRY IT: *Using all three motors (all on the same axle), make a high-speed drag racer. Use the gears in your kit to further increase the axle speed by making larger gears rotate smaller gears.*

WARNING: *The drag racer can attain high speeds. You are responsible for any damage to your NXT parts (such as gears snapping) or injuring someone. Be extremely careful.*

Cables

A: Short Cable   (1 black)   55804
B: Medium Cable (4 black)   55805
C: Long Cable   (2 black)   55806

Standard MINDSTORMS cable lengths are 20 cm (8.5"), 35 cm (14.5") and 50 cm (18.5"). One thing not immediately obvious is that if the cable is laid flat, the RJ12 connector faces up on one end and down on the other. This so the wires connect to the correct pins on the RJ12 connectors.

There is no part to connect two RJ12 connectors together into one longer-wire. However, different lengths can be purchased from other vendors for a modest cost. HiTechnic supplies a set of four wires: 12 cm (4.75")–extra short, 16 cm (6.3")–short, 70 cm (27.6")–long, and 90 cm (35.4")–extra long. Mindsensors.com sells wires that use ribbon cable. The advantage of ribbon cable is that they flex and twist easier than hard rubberized cables; the downside is they are not as rugged.

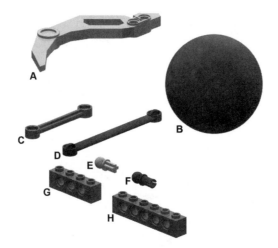

Other Parts

A:	Bionicle Weapon Pincer Suukorak	(4 silver)	50914
B:	Ball 5.2 cm	(1 blue, 1 red)	41250
C:	Steering Link Type 2	(3 grey)	2739B
D:	Steering Link 9L	(2 black)	32293
E:	Axle Towball	(2 grey)	2736
F:	Friction Pin with Towball	(8 black)	6628
G:	Brick 1 x 4 with Holes	(4 dark grey)	3701
H:	Brick 1 x 6 with Holes	(2 dark grey)	3894

The Bionicle pincers are useful for claws, as demonstrated by Tribot. The ball is mostly used as an object for your robots to manipulate. However, some builders have used the ball (along with the pincers) as a caster wheel.

Steering links and towballs are useful for steering mechanisms and tilting. *Alpha-Rex* uses these parts to shift the robot from side to side when it walks, and to move the arms and head.

The next chapter will examine different ways to use collections of parts from this chapter.

Building 101

Topics in this Chapter

- Design Patterns
- LEGO Laws
- Engineering Goals

Chapter 6

I n the previous chapter we explored the individual parts contained in the NXT kit and identified their function. We also described the versatility of the parts and how they can be used for more than one purpose. In this chapter we will look at collections of parts; mechanisms that consist of two or more parts but produce a single function. We will also look at some of the principles that govern the practical use of LEGO. The final part of this chapter will explore general goals that apply not only to LEGO, but to any design.

Design Patterns

Design patterns are ways in which parts are put together to produce a specific function or mechanism. While much of the great challenge and fun of LEGO is discovering how to create mechanisms, there is no need to re-invent the wheel. This section will help you to familiarize yourself with common design patterns to give you a head start on your projects.

WEBSITE: Nxtasy.org *contains a section with LEGO mechanisms. Go to* www.nxtasy.org *and click on Projects.*

Caster Wheels

Two-wheeled robots are the most common platform for robotics because they are ideal for navigating in confined spaces. These types of robots usually require a smaller third wheel, called a *caster wheel*, to maintain balance. A caster wheel has the property of being able to rotate in all directions. You can see caster wheels on rolling desk chairs.

The LEGO NXT kit only includes large wheels, but in general, you should use a smaller caster wheel than the drive wheels since they swivel more easily than large, wide tires. In the NXT kit, you can substitute a double-bevel gear or the pulley wheel for a castor (see Figure 6-1).

Figure 6-1 A caster wheel

The placement of the caster on your model is important. Casters form the third point of a triangle in your model, with the two drive-wheels forming the other two points. This gives your robot stability. Casters work best when they are in the same circle as the main wheels (see Figure 6-2 A). If you notice your vehicle jumps around while rotating, the caster is probably too far outside the circle (Figure 6-2 B). You should also position the mass of your robot over the center of the triangle for stability.

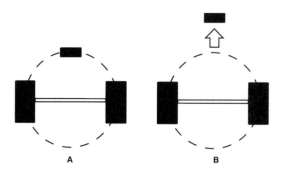

A   B

Figure 6-2 Proper and improper placement of caster wheel

Lateral Motion

Lateral motion is movement in a straight line. Some devices that require lateral motion are scanners and photocopiers. Normally these machines use a gear rack, which is a beam with many gear teeth. NXT kits do not contain gear racks, requiring that you find an alternate method for producing lateral movement.

Lateral motion can be difficult to achieve because a motor moves in a circle. Technically a vehicle achieves lateral motion by driving forward. This solution was used for the scanners in this book. However, there is a way to produce constrained, lateral motion without using bulky tires.

Figure 6-3 Lateral motion using a worm screw

Figure 6-3 demonstrates a mechanism using NXT parts to create lateral motion. It relies on a worm screw and a half-bush on an axle. The worm screw engages the half-bush and moves it a small distance. To achieve greater lateral movement, more half-bushes can be added to a longer extended axle.

Periodic Movement

Sometimes you want a robot to perform novel periodic movement (such as turning its head or moving arms) without using an extra motor. You can create periodic movement by placing a gear on a main drive axle using one of the off-center holes on the gear (see Figure 6-4). The off-center gear borrows torque from the drive axle, periodically rotating another gear at intervals thereby generating periodic movement.

Figure 6-4 Movement at intervals with an off-center gear

You can also use a variation of this mechanism to sequentially rotate four axles one after another using only one motor! This can be used to move four legs in sequence. To accomplish this, place up to four gears around the off-center gear.

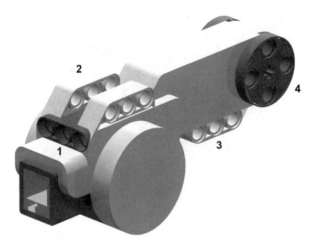

Figure 6-5 NXT motor anchor points

Attaching Motors

Sensors are very easy to attach, whereas motors seem to pose a greater challenge. But, don't be fooled! LEGO NXT motors have three basic anchor points, plus the axle hole which can also serve as an anchor point (see Figure 6-5). Surprisingly, motors don't need an especially secure attachment; one pin at either of the connection points, plus the axle seems to be good enough in many cases.

The unusual shape of the motors may also appear to be a hindrance, but the design actually works well in most types of robots. For example, when building a robot arm, the motor extends the length of the arm. As a result, you can build robots using fewer parts than with the old RIS kit.

Often a design requires two or more motors joined together. This is easily accomplished by inserting a beam between the two motors to connect them (Figure 6-6A). Two motors can also be connected without additional beams (see Figure 6-6B).

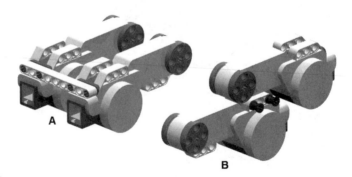

Figure 6-6 Interconnecting motors

Figure 6-7 Gimbals joint from two motors

The orange axle-hub of the motor is very versatile, and can accommodate not only axles, but other parts. For example, you can connect one motor to another motor perpendicularly to produce a *gimbals joint* (Figure 6-7). A gimbals joint is able to rotate along all axes acting much like your shoulder joint.

Sometimes when you attach your motor, the forward function makes the motor rotate backwards. So which way is forward? Figure 6-8 indicates forward movement of the motor. This is not necessarily a problem, since you can easily alter your code.

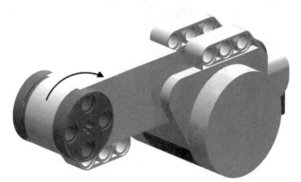

Figure 6-8 The default forward direction

TIP: *When building a robot, you might periodically want to test movement without writing a small test program. A simple method to do this is to remove one gear of the gear-train so the motor is no longer resisting, and then spin the part with your fingers. Using this method, you can gauge how hard the motor must work, which can get you focused on making low resistance gear ratios.*

Two Axles with One Motor

Under some circumstances, multiple motors are required. A potential limitation can occur when a fourth motor is required since only three motor ports are available. To circumvent this problem, you can create a virtual motor using a worm gear to split the movement of a single motor.

The worm gear can slide up and down an axle. Using this property, it is possible to make one motor control two different axles at different times (like having two single direction motors instead of one multidirectional motor) thereby reducing the number of motors needed. Figure 6-9A shows this mechanism.

You can also alter this design to control one axle in both directions and the other axle in one direction only. Simply placing a bush on one side of the worm-screw axle will produce this effect (See Fig 6-9B).

A   **B**

Figure 6-9 Dual movement from one motor

Differential

The differential is an important component in the steering mechanism of vehicles. When a car turns a corner, the inside wheels rotate less than the outer wheels. If both drive wheels are on the same axle, they are forced to turn at the same rate, which can cause the wheels to skid and bounce. The differential splits one axle into two, and allows torque to be distributed to both axles to prevent skidding.

The NXT kit does not contain a differential. But, by following the six simple steps detailed below, you can build your own differential using NXT parts!

STEP 1 Add parts as shown.

STEP 2 Add parts as shown.

STEP 3 The 4-unit axles join the long pins together.

STEP 4 Add parts as shown.

STEP 5 Add parts as shown.

STEP 6 Add parts as shown.

Cable Management

Sometimes it can be heartbreaking to have to plug cables into a clean robot design. Suddenly your neat robot looks like a Medusas head of tangled wires. To add to your troubles, the cables are semi rigid, not pliable like a rubber band. Due to this rigidity, they can interfere with the movement of your robot. When this happens, you need to think seriously about cable management.

If you built LEGO's Alpha-Rex model, you understand the challenges involved with managing cables.The designers of Alpha-Rex even included special pins that protruded from the robot specifically for cables.

The easiest method of cable management is to allow the cable to spring out from the back of the robot in an arc, clear of any obstructions. This gives your robot a Japanese Manga look. This also protects the NXT brick, since the wires can cushion the impact if your robot falls over.

If the cables continue to get in the way, the standard means to remedy this is to use two long pins with bush (see Figure 6-10 A). To trap three cables, use long-pins spaced one unit apart with a 3-unit beam to cap it off (see Figure 6-10 B).

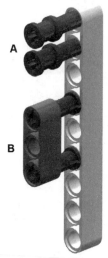

Figure 6-10 Cable management

Sometimes the selection of cables in the NXT kit is not appropriate for your design. In this case, you can purchase different lengths from LEGO, Hi-TECHNIC or Minsensors.com (see Appendix A).

LEGO Laws

When people think of LEGO, what often comes to mind is little square bricks with studs that lock onto other bricks. Upon opening an NXT kit, these people may wonder what happened to the LEGO. They might be hard pressed to find any bricks with studs.

NXT kits contain a menagerie of unusual parts that give LEGO a whole new look. Whereas the old yellow RCX brick had studs all over it, the NXT unit and sensors only have holes for pins and axles. There is a new paradigm of building without studs, and this section will get you acquainted with this new approach.

When you are faced with limitations, problems are solved by being creative. For example, creativity seems to flow in constructing a Haiku when one is limited to only a few syllables. Likewise, LEGO parts are simple structures with limits to their application. You are restricted by the number of parts, the type of materials, or how the parts can be attached to one another. Rather than impeding creativity, I think these limitations enhance creativity.

Bricks are good for building houses, but in the real world bricks are not used to build robots. You need rotating axles and structural beams for most robot projects. For this reason, the emphasis in the NXT kit is on beams.

Before we look at specifics, you might wonder if the studless paradigm is better. Figure 6-11 shows a LEGO arm that uses traditional studded construction next to an NXT arm featured in this book. The old method required 145 parts to build (including two rotation sensors), while the NXT arm only required 121 parts. The NXT even performed better. With certain designs, therefore, studless NXT robots require fewer parts to perform the same task.

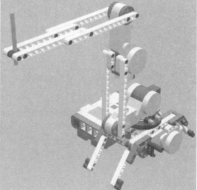

Figure 6-11 Studded vs. studless construction

There is also style to consider. The old RIS arm looks like something from the 1980's while the new LEGO NXT arm truly embodies what you would expect from the second millennium. It's no accident that the LEGO Star Wars kits, such as the Star Wars Hailfire Droid and the Destroyer Droid, use studless construction. It simply looks more modern.

In the world of engineering, machines require screws, nuts, bolts, nails, glue, or welding to hold parts together. LEGO does not use these types of joints. Instead, everything is held together by friction. Friction, however, does not produce the most secure joint, especially when large forces are subjected to your models by powerful servo motors. Let's explore how to build functional models that won't fall apart.

Take a good look at the pins in the previous chapter. These are the main components for holding your robot together. The black friction pins are especially plentiful in the kit because these are the most versatile connectors.

The three-hole structure is everywhere on the main NXT parts— on sensors, motors, and the NXT brick. This is the key to understanding how to connect different parts. The complements to the three-hole structures are the bent beam with four-pins, or the perpendicular joiner (both light-grey). Under most circumstances, you should use these parts when connecting a beam to the NXT brick in a normal or perpendicular orientation.

At first, you may feel that you need precognitive abilities to predict whether or not the beam-holes are going to line up. Rest assured that LEGO knows that we don't all have extrasensory powers. They have designed these parts so that beams and holes will almost always line up at regular intervals.

Building is exciting. There are times when you will have your model working, yet you see a way to make it better. In order to make that change you need to rip apart what you already have and risk losing a good design in the process. But, you are driven to take the chance to turn a good design into a great design. It's a scary moment that can pay off with success or lead to frustration.

 TIP: *A digital camera can be your best friend when you are building. Take a photograph of your robot if you are about to try out something new that requires disassembling a large part of your robot. If something goes wrong you can always use the photograph to help you rebuild your original design.*

Stacking Beams

To stack studded bricks, you simply press the bottom of one brick into the top of another to gain elevation. With studless design the solution is not as obvious.

Perpendicular attachment to gain elevation, as shown in figure 6-12A, is not structurally stable and will wobble. To create a stable structure as depicted in figure 6-12B, you can use L-beams, although this approach can be somewhat bulky. Another option is to attach axle joiners perpendicularly to the main beams. I favor these parts when I need to stack one beam on top of another.

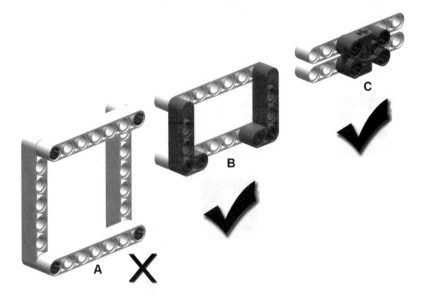

Figure 6-12 Stacking beams

TIP: *To avoid* LEGO *finger (a harmless reddening of your fingertips) avoid pulling on small parts when trying to remove them from the structure. Axle-pins tend to be the worst offenders. Using an axle to push parts out from the other side will make it easier to pull small items out.*

Changing the Axis of Beams

Often when building a robot, you need a row of holes to line up along a beam but they are in a different orientation. This prevents you from attaching an axle or another beam along the proper line (one of the weaker aspects of building with LEGO). To overcome this problem you must transform the orientation from one axis to another – for example, from the x-axis to the y-axis.

There are several ways to do this. Figure 6-13 shows three beams with holes aligned along all three axis. The perpendicular axle joiners provide a sturdy connection at a 90-degree angle. Simply insert a 2-axle to connect the complimentary axle joiners then attach them to the beam with friction pins.

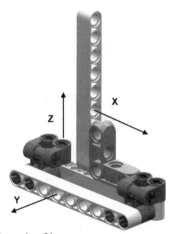

Figure 6-13 Changing the axis of beams

There are other parts that can change the axis of beams, such as the dual perpendicular joiner (see Figure 6-14). However, because this part is symmetrical along the middle of the holes, it can change the spacing, and the holes along the new beam may not line up with the rest of your model.

Figure 6-14 Other methods for changing the axis of beams

Securing Gears

Although it isn't a good practice, it is possible to get away with poor design in some aspects of robot construction. Gears however, do not tolerate flimsy design. They need to be properly supported or they will slip while under torque.

Gears that are not properly supported make a telltale clicking sound. This occurs when two gears push away from each other, likely because they are on different beams. When gears are not on the same beam (see Figure 6-15A) the beams will flex, causing the gears to slip. By having two gears on the same beam, this does not occur and the gears function properly (see Figure 6-15B).

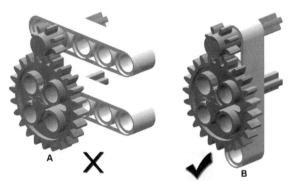

Figure 6-15 Improper and proper gear support

It's also important to position the gear right next to the beam. If the gear hangs out from the beam, it can slip (see Figure 6-16A). You can force the gears to hug each other by adding a small beam to the ends of each axle (see Figure 6-16B). Even though the gears are hanging precariously in the open, they still work well.

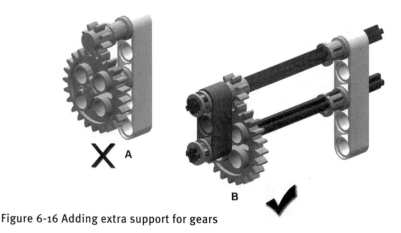

Figure 6-16 Adding extra support for gears

Engineering Goals

Design goals that apply to the real world of engineering also apply to LEGO: LEGO, after all, also operates in the real world. When designing machines, there are steps that can help you refine your creation. In engineering, this process is called *requirements analysis*. With this method, you start with the requirements and design around them. This section will take a brief look at how requirements analysis can help you.

1. Define Functionality

It's important to know the primary function of your robot and focus on that. Ask yourself, what is the function of this robot? Sometimes it's merely to drive from one point to another. In this case, you need to determine the best steering method for your robot. Does it have to turn on the spot? Is analogue steering more useful?

2. Design around functionality

Once you've determined the functions of your robot, identify the most important part of your robot and build it first. For example, if you are making a robot arm, the claw is probably the most important feature, since every other part of the robot is in the service of moving the claw. Once the claw is done, you can add the supporting mechanisms, such as the arm that supports the claw. When the arm is done, build a base to support the arm.

3. Common design goals

Goals give you something to strive for. Sometimes the list of goals isn't apparent until you've played with the parts and constructed a prototype. For example, with arms, I didn't realize one of my goals was to make the arm move fast until I tried it out. Similarly, only after trying to have it lift an object did I realize one of my goals was to have it to lift objects of some mass. Ironically, sometimes you realize what all your goals were only when your project is done.

To help you pick your goals, think of the following properties most engineers want from their creations:

- Speed
- Strength
- Power
- Agility
- Stability
- Accuracy
- Symmetry
- Compactness
- Minimalism
- Robustness
- Modularity

How fast do you want it to move? What sort of payload might your robot carry (arm or vehicle)? Is your robot going to race with other robots? Will it fight with other robots? Depending on the answers to these questions, you will have to provide your robot with the appropriate properties to attain your goals.

Strength

Strength is always valued, even if your robot won't experience rough use. Try dropping your robot from a height of 10 centimeters onto carpet and note what breaks off. Keep working on those joints until they don't break anymore. Some robots break easily when you try to pick them up, so try picking it up in different ways. Study the joint that failed and develop a means to strengthen that joint until it holds tight.

Speed

Speed is achieved in two ways: motor/gear/axle speeds and programming. The ARM processor inside the NXT brick is fast. It processes faster than the motors are able to react. Take advantage of this. Don't have your robot pause using Thread.sleep() unless it is absolutely necessary. Try upping Motor.setSpeed()as high as it can go. See if the LEGO motors can handle higher gearing, which can make your axle spin faster.

Power

The tradeoff with speed is power. Since the specifications of LEGO NXT motors are the same, you can have speed or power, but not necessarily both. A powerful robot with geared-down axles will be slower. You decide which is more important for your creation.

Minimalism

Believe it or not, you want to use as few parts as possible. There's a saying in engineering, "You are done when you can no longer remove any parts". For starters, make sure there are no unattached pins sticking out. But, more importantly, optimize your design to use as few pieces as possible without compromising the essential features of your robot. For example, if your design uses four beams and you can come up with a design that does the same job with only three beams, the latter is superior.

Compactness

Making your robot mechanisms as compact as possible for the size of your robot. That is not to say you should make your robots small; it is perfectly valid to want or require a large robot. However, the mechanisms like gearing should be as compact as possible. Allow as little air as possible between parts.

Stability

Stability is important to any design. Whether it is a stationary device or a moving robot, your creation should not be prone to tipping over. Chapter 11 on balance and walking explores this concept in more detail.

Robustness

A robust robot is not only good on hard floors but also carpet. Your robot is robust when it deals well with diverse conditions. Making your robot easy to tune can also increase its robustness. For example, a robot that allows you to adjust the length of the legs can handle diverse situations.

Accuracy

An arm that is only accurate to within five centimeters might not be very effective if it tries to pick up a marble. You can increase accuracy by gearing down motor movements. Also, make sure your gears don't slip, otherwise your motor tachometers will no longer provide useful data.

Modularity

Modularity is important, especially with complex machines. The robot arm in this book is a prime example of modularity. It consists of a base unit that rotates like a crane, a lower arm, a forearm, and a claw. If you want to change any part of this arm, you can change one module without having to reconfigure the other modules.

To make modularity work, you need to know how the modules connect to each other. In practice, modularity is often inherent when you build something because modules tend to group together naturally.

Bite into Bluetooth

Topics in this Chapter

- Introduction to Bluetooth
- iCommand
- Moon Buggy – a radio controlled vehicle

Chapter 7

Bluetooth is a major feature of NXT. I have always been excited about wireless control of robots because of the enormous resources and brainpower that can be harnessed through wireless communication. LEGO tried wireless control previously with the green Cybermaster brick. Unfortunately, because of several limitations, it was not as successful as the RCX. Did LEGO get it right this time with Bluetooth? This chapter will answer that question.

Meet Bluetooth

Bluetooth is a wireless protocol that allows devices to communicate with each other. The most popular Bluetooth devices include mice, keyboards, wireless headsets, and PDA/cell phones. In fact, a Bluetooth equipped cellular phone or PDA can even control your NXT brick. Likewise, your NXT can control Bluetooth devices, including other NXT bricks.

Bluetooth was introduced in 1998 and was initially slow to catch on, but vendors are now rapidly adopting it. Both Microsoft and Logitech manufacture Bluetooth Wireless Keyboard and Mouse combos. Wireless headsets for cellular phones are becoming ubiquitous. In 2006, both Sony and Nintendo released video game consoles that use Bluetooth. NXT has also given a boost to Bluetooth.

But what is Bluetooth? The technology can be compared to USB in many ways. A single Bluetooth dongle effectively acts like many wireless USB ports. Bluetooth eliminates cables from keyboards, mice and game controllers.

The major advantage of Bluetooth versus the infrared (IR) system used by the RCX is the elimination of line-of-sight problems. Radio waves travel through solid objects and in all directions, whereas the IR port on the RCX had to point right at the IR tower with no intervening structures.

Many people wonder how Bluetooth is different from Wi-Fi (the 802.11 standard). For starters, Wi-Fi is designed for networking, usually with one or more wireless routers and many computers. You wouldn't imagine using an 802.11g network to control a mouse wirelessly—it would be unresponsive and costly. Bluetooth is for data communications between two devices only – usually a small device and a personal computer.

NOTE: Bluetooth operates on 2.45 GHz, while WAP 802.11g is on 2.4 GHz (see Table 7-1). Although these frequency bands overlap, you probably won't experience conflicts; Bluetooth and Wi-Fi use frequency hopping, jumping from one band to another up to 1600 times per second.

The only time my Bluetooth connection appeared unstable was in an area with five wireless networks (including the one connected) plus a cordless phone system. While I was streaming music through the network the keyboard seemed slow and unresponsive.

Device	Frequency
Bluetooth	2.45 GHz
Wi-Fi (802.11g)	2.4 GHz
Cordless Phone	900 MHz (older), 2.4 GHz, or 5.8 GHz
X-10 Camera	2.4 GHz
Cell Phones	800 MHz
Headphones	924 - 928 MHz

Table 7-1 Wireless frequencies of common devices

Bluetooth is cheaper and it uses less power than Wi-Fi. This is important for a device like the NXT since Wi-Fi would quickly eat up those batteries. Bluetooth typically consumes about 1/100th of the power of Wi-Fi (802.11b).

Bluetooth is capable of transmitting data at 460.8 Kbit/s, which is on the low end of Wi-Fi devices (see table 7-2). However, these speeds are more than adequate for transmitting programs and files to the NXT brick. You will notice a huge improvement in speed compared to uploading programs to the RCX.

Wireless Standard	Typical Data Rate
RCX IR Port	2.4 Kbit/s
Bluetooth	460.8 Kbit/s
Wi-Fi 802.11b	6.5 Mbit/s
Wi-Fi 802.11g	24 Mbit/s
Wi-Fi 802.11h	200 Mbit/s

Table 7-2 Comparing wireless speeds

Multiple Bluetooth dongles can be used in the same area since Bluetooth uses frequency hopping to avoid conflicts. There shouldn't be much of a problem in a classroom setting, even with dozens of Bluetooth dongles. If interference does occur, USB cables can be used to transmit code to robots

Pairing too many devices to the same Bluetooth dongle is a larger problem. I've found that other Bluetooth devices are affected when there is a lot of traffic over a single Bluetooth connection, voice communications for example. My Bluetooth mouse behaved like a drunk when I was using a Bluetooth headset.

Not all Bluetooth dongles are compatible with the NXT. First, it must be Bluetooth 2.0 or higher. Second, under Windows it must support the Widcomm® Bluetooth stack. Some Bluetooth solutions, such as those built into Dell notebook computers, are not compatible. If you aren't sure and don't want to go through the trouble of researching this, you can order a Bluetooth dongle from LEGO (see Appendix A). As long as Linux users can use their dongle with the BlueZ Bluetooth stack, there should be no compatibility problems.

TRY IT: *If you want to confirm that your dongle is using the Widcomm Bluetooth stack, do the following:*

1.1 Select Start > Control Panel > System.

1.2 Select the Hardware tab, then click the Device Manager button.

1.3 You will see a list of devices, including Bluetooth. Expand the Bluetooth selection and highlight the Bluetooth wireless hub (see Figure 7-1).

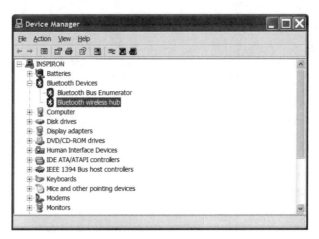

Figure 7-1 Bluetooth devices in Windows

1.4 Click the Properties icon and you will see some general information. Click the driver tab to view the device driver (see Figure 7-2).

Figure 7-2 Device driver information

1.5 As long as you see Widcomm and any version higher than 1.4.2.10 you should have no compatibility problems.

Pairing your NXT to your PC

Pairing occurs when one device tells another device that they can be friends. The reason for pairing is to avoid someone accessing your Bluetooth devices without permission. By pairing, you give the devices permission to interact (like plugging a USB cable into a computer).

Pairing is done by locating the Bluetooth device, then entering a four digit combination on both devices. I recommend using the LEGO software to pair the devices. If there are any problems, the software will attempt to help you.

TRY IT: *If for some reason you don't want to use the LEGO software, you can pair the devices through your operating system.*

1.1 Turn on your NXT brick. From the main menu, select Bluetooth.

1.2 Hit the orange button to select On/Off so it says On. Hit the orange button again.

1.3 Go into Bluetooth again, select Visibility. Select Visible.

1.4 Hit the grey escape button to go back, then select search.

1.5 Eventually it will come up with your computer name if the Bluetooth Dongle is installed correctly. Select your computer.

1.6 After a while your computer will come up with a window asking for a code. Type 1234, or make up your own.

1.7 The NXT brick will also ask for a code. Type the same code. The two devices are now paired.

Introducing iCommand

The leJOS NXJ project has a sister-project called iCommand, a package for remote control from a computer. This package seeks to mirror leJOS NXJ as closely as possible, which means the classes and methods tend to look similar. The major difference is that leJOS NXJ code runs on the NXT brick while iCommand code runs on your computer. It controls the NXT brick by sending individual commands wirelessly. Whether you choose to use leJOS NXJ or iCommand depends on the requirements of your robot. Let's compare leJOS NXT with iCommand.

iCommand Advantages:
- Access to all devices on your computer (see Figure 7-3)
- Access to memory on your computer.
- More powerful display for output (100 x 64 mono vs. >800 x 600 full color)

iCommand Disadvantages:
- Not as fast as leJOS
- Limited by what LEGO protocol allows (e.g. can't draw on LCD screen)
- Not as portable if you want to take your robot somewhere.

Speed is often a concern for robot programmers. Because all commands are sent to the NXT through Bluetooth, iCommand reacts a little slower than code on the NXT brick. However, most robots are slow and don't require fast processors. It takes a few seconds to move

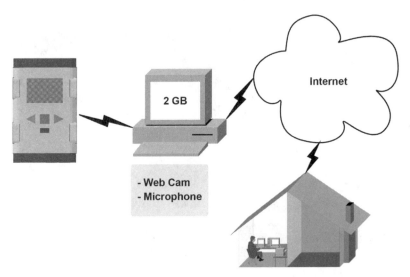

Figure 7-3 Exploiting your computer's resources

a foot, so a hundred milliseconds here and there isn't a big deal for most robot applications.

Since iCommand is running on your computer and uses the standard Sun Java API, you have access to an incredible number of resources, such as the Internet and other hardware devices on your computer. Your imagination can go wild thinking of interesting new creations.

TIP: It's easier to troubleshoot iCommand programs using the console to output messages. NXT programming can be a little more difficult to output to the LCD screen. Even if you are writing for leJOS, you might want to write your initial code under iCommand and then change it to leJOS after the bugs are worked out.

Installing iCommand

The steps below will show you how to install iCommand on your computer.

NOTE: If you prefer using an IDE to develop iCommand code, I recommend Eclipse. Start a new folder just for your iCommand projects (separate from your leJOS projects). The steps below also show you how to set the classpath within Eclipse.

1. Make sure you have the latest version of the Java Development Kit installed. If you installed leJOS, you already have this installed.
2. Download iCommand from: www.lejos.org
3. Unzip the contents of the zip file into a directory.

4. Set your classpath to include lib/icommand.jar. In Eclipse, create a new project, then select Project > Properties > Java Build Path > Add External Jars... and browse to icommand.jar.

Next we need to install a program that allows serial communications with Java. Only Windows and Macintosh users need to install this as Linux users can use BlueZ directly. Installation is slightly different for Windows and Macintosh users.

RXTX Install – Windows

1. Download RXTX from www.rxtx.org and extract the contents of the download.

2. Set the Java classpath to include RXTXcomm.jar.

3. Optional for Eclipse only: Start a new project just for Eclipse code. Then select Project, Properties, Java Build Path and "Add External Jars...", then browse to RXTXcomm.jar.

4. There are some dlls that come with RXTX in \Windows\i368-mingw32. Make the dlls accessible by copying them to your Java bin directory e.g. c:\j2sdk1.4.2_12\bin.

5. Optional for Eclipse only: Select the RXTXcomm.jar you just added and expand it by clicking + (see Figure 7-4). Select *Native library location* and browse to the RXTX subdirectory \Windows\i368-mingw32.

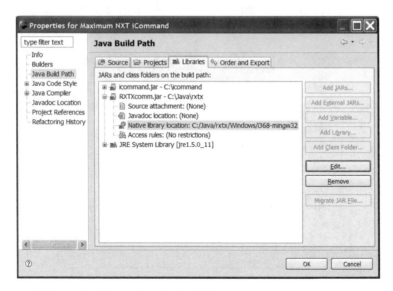

Figure 7-4 The native library location in Eclipse

RXTX Install – Mac OSX

1. Download RXTX 2.1 or higher from www.rxtx.org.
2. Extract the zip file.
3. Copy files RXTXComm.jar, rxtxSerial.jnilib and iCommand.jar to the directory /Library/Java/Extensions/ so you don't have to set the classpath.
4. Navigate to MACOSX_IDE\ForPackageMaker\
   Run the RXTX_Tiger package and follow the instructions.
5. Make a directory called /var/lock and give it read/write permissions (777).
6. Alternate: Some users create a directory /var/spool/uucp and set up appropriate permissions (777). Include your username in the uucp group using the NetInfo manager.

Now we need to set up a communications port in order to associate it with Bluetooth. This is different for all three platforms.

Setting the COM port – Windows

When you installed the LEGO software, it set up a COM port automatically. You just need to figure out how to find the COM port.

1. Go to Control Panel > System. Click the Hardware tab and select Device Manager.
2. Expand Ports, as shown in Figure 7-5. You need to choose one of these ports for iCommand. I picked the lowest one, but you can actually pick any one, such as COM4.

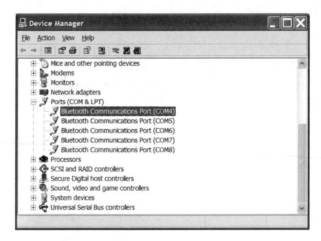

Figure 7-5 Selecting the Bluetooth port

3. iCommand came with a file called icommand.properties. Set the value of nxtcomm to the value of your COM port, e.g. COM4.

4. Put the file icommand.properties into your home directory (C:\Documents and Settings\user) or working directory (where you compile your code). I recommend the home directory.

Setting up the Serial Service – Mac OSX

1. Pair your NXT and your computer with the Bluetooth setup assistant.

2. Go to *Open Bluetooth Preferences,* select *Edit Serial Ports* for your NXT and name it. This should create a new serial port, /dev/tty. NXTPORTNAME

3. In the file icommand.properties that came with iCommand, set the value of nxtcomm to the value of your serial port.

4. Put the file icommand.properties into your home directory, or your working directory.

5. To check the location and contents of the file, execute the command: java -jar icommand.jar

BlueZ Install – Linux

The default installation of iCommand makes use of RXTX to set up a Bluetooth link with the NXT brick. For the Linux platform there is an alternative installation method that makes use of a native library to interface directly with the Linux BlueZ Bluetooth protocol stack (www. bluez.org).

1. Your Linux platform must have the required packages installed to support the BlueZ Bluetooth protocol stack. Please consult the documentation of your Linux distribution about how to install BlueZ. In general, this means installing bluez-utils, bluez-libs, bluez-gnome, and a kernel with Bluetooth support and the bleutooth kernel modules.

    Ubuntu already contains all the required Bluez packages, except for the *bluez-passkey-gnome* package. This package is required to exchange a passkey while pairing the NXT brick with the Linux platform.

2. The native library 'libicmdbluez.so' to interface with BlueZ is provided with the iCommand distribution. Copy the library into the Java Development Kit (JDK):

    ```
    cp libicmdbluez.so $JAVA_HOME/jre/lib/i386
    ```

3. Install the iCommand java library 'icommand.jar'. Copy the java library 'icommand.jar' into the JDK:

    ```
    cp icommand.jar $JAVA_HOME/jre/lib/ext
    ```

4. Insert the Bluetooth dongle into the USB port. It is actually better to boot Linux with the Bluetooth dongle already inserted.

5. Pair your NXT brick and the Linux PC. Initiate the pairing process from the NXT brick. Send the default passkey '1234' from the NXT

brick. On your Linux PC, answer the incoming pairing request with the same passkey '1234'.

6. We now need to find the Bluetooth address of your NXT brick. Scan the environment for Bluetooth devices. Perform the command:

```
hcitool scan
```

It will produce something like the following:

```
Scanning ...
    00:14:A7:6D:C6:AA      Nokia 6230i
    00:16:53:01:EC:04      NXT
```

The output shows the Bluetooth address and the friendly name of the discovered Bluetooth devices. Note the Bluetooth address of your NXT brick. In the sample output this is: 00:16:53:01:EC:04

7. Configure iCommand in the file icommand.properties:

   a. Set the property nxtcomm.type to the value 'bluez':

   ```
   nxtcomm.type=bluez
   ```

   b. Set the property bluez.address to the Bluetooth address of your NXT brick:

   ```
   bluez.address=00:16:53:01:EC:04
   ```

   c. Comment out all other lines using the character #.

   d. Place the file icommand.properties in the home directory of your UNIX account.

Testing iCommand

That's it. You are now ready to use iCommand. You can use either the standard LEGO firmware or the leJOS replacement firmware. iCommand works with both because it uses the LEGO Communications Protocol (LCP).

There are several sample programs included with iCommand. To test these, compile and run them as normal Java programs. They will compile and execute within Eclipse without any special development tools. If you are using the command line to compile and run Java code, use the regular JAVAC and JAVA commands. For example, in the samples directory is a program called Beep:

```
javac Beep.java
```

Now run the program:

```
java Beep
```

If the speaker of the NXT brick produces a beep sound, then everything is installed and working properly.

Installing iCommand on a Pocket PC

You can use iCommand on a PDA or mobile phone using Windows Mobile (2003 or 5.0). Upload code, jar files and classes to your Pocket PC via USB using Microsoft ActiveSync.

1. Download and install the Mysaifu JVM:
    www2s.biglobe.ne.jp/~dat/java/project/jvm/download_en.html
    Install by copying the cab file to your Pocket PC and tapping it.

2. Download the Java Communications package for Mysaifu:
    http://prdownloads.sourceforge jp/mysaifujvm/16920/
    javaxcomm.0.0.1.zip

3. Copy javaxcomm.jar to \Program Files\Mysaifu JVM\jre\lib\ext on your Pocket PC. (You might need to create the ext directory.)

4. Copy javaxcomm.dll to \Program Files\Mysaifu JVM\jre\bin on your Pocket PC.

Figure 7-6 Establishing a Bluetooth connection

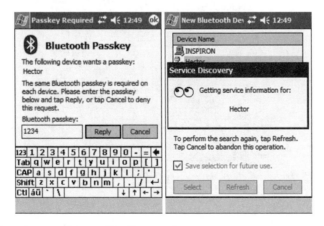

Figure 7-7 Entering passkey and finding services

5. Establish the Bluetooth Connection between the PDA and NXT. Look in Bluetooth for devices (see Figure 7-6). You should see your NXT name in the list (Hector in my case). Click it, then click select, and it will prompt you for the passkey if they are not already paired (see Figure 7-7).

6. Check which port Bluetooth is using. Tap the Bluetooth icon, and then go to Bluetooth Settings > Tab Services > select Serial Port > press Advanced. Note the value of *Outbound COM PORT.*

7. Edit the iCommand file icommand.properties and set the value of your COM port followed by a colon (:). It is important to include the colon, otherwise it will not work.

   e.g. COM6:

8. Modify the icommand.properties file to use the Sun Java Communication API instead of RXTX. The complete file should appear as follows:

```
nxtcomm=COM6:
nxtcomm.type=sun
```

9. Transfer icommand.properties to the \My Documents directory on the Pocket PC.

10. Copy icommand.jar to \Program Files\Mysaifu JVM\jre\lib\

11. Add icommand.jar to the Mysaifu class path (see Figure 7-8). First click Options on the main screen, then add icommand.jar to the classpath using the keyboard:

    ;\Program Files\Mysaifu JVM\jre\lib\icommand.jar

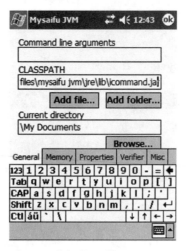

Figure 7-8 Setting the class path in Mysaifu

12. Now you can run your own iCommand applications on the Pocket PC. Copy a compiled iCommand class file to the folder My Documents on your Pocket PC.

13. Now run your application from Mysaifu. If your code uses System. out, make sure to check *Show console* (see Figure 7-9).

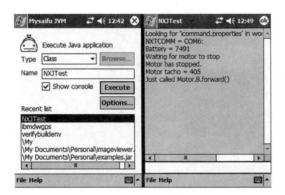

Figure 7-9 Choosing a class to run

 WARNING: *Since Mysaifu is not a complete Java API, some of the iCommand classes that use a file system will not work properly. The majority of the classes work well, however.*

iCommand API

The iCommand API is similar to leJOS NXJ, in fact, it is almost identical. In some places such as the navigator, behavior and sensor classes, the source code is the same. It would be redundant to list the entire iCommand API, so let's instead focus on the differences.

 NOTE: *Motor rotations are more accurate when you use the leJOS firmware as opposed to the standard LEGO firmware. If you decide to use iCommand with the LEGO firmware, make sure to upgrade to the latest firmware.*

icommand.nxt.comm.NXTCommand

The most important difference between iCommand and leJOS is that iCommand won't work until you call the open() method. When you are done your program should call the close() method. The following example is typical of iCommand code. It looks like leJOS NXJ code except for the import, open() and close() methods:

```
1. import icommand.nxt.comm.NXTCommand;
2. import icommand.nxt.*;
3. public class Test {
4.
5.     public static void main(String[] args) {
6.         NXTCommand.open();
7.
8.         System.out.println("Battery = " + Battery.
            getVoltageMilliVolt());
9.
10.        UltrasonicSensor us = new UltrasonicSensor(Sensor
            Port.S3);
11.        System.out.println("Distance " + us.getDistance());
12.
13.        Motor.A.rotate(1000);
14.
15.        System.out.println("Closing shop");
16.        NXTCommand.close();
17.     }
18. }
```

Another option is setVerify(). When this is set to true, every command sent to the NXT is verified to make sure it worked. When setVerify() is activated, calls like Motor.forward() return a value from the NXT. If this value is something other than 0, something went wrong.

I've found no real reason to use setVerify(). iCommand is reliable and the commands don't usually fail unless the methods are somehow handled incorrectly. Unless your robot is engaging in some life-threatening task, such as surgery, then you probably won't need to use setVerify(). It just slows down the operation, taking roughly 100 ms longer per command.

icommand.vision

The vision package is a major departure from leJOS NXJ. This package allows you to use video data with your robots. Since the package runs on your PC, there is no complement for leJOS NXJ. Chapter 10 will explore this package in detail.

A Radio Controlled Car

One of the classic uses of radio communications is the radio controlled car, or in this case, a radio controlled moon buggy. This little turbo charged beauty uses analog steering, unlike the other mobile robots in this book. I call it turbo charged because the gearing speeds up motor rotation about threefold.

Ideally, this vehicle would use a differential for driving the rear axle. Without the differential, both rear tires rotate at the same rate when the vehicle is turning. Something has to give, so one of the tires will skid. This results in a poor turning radius. With a differential along the rear axle, the turning radius would increase dramatically. The NXT kit does not contain one, so we will do without.

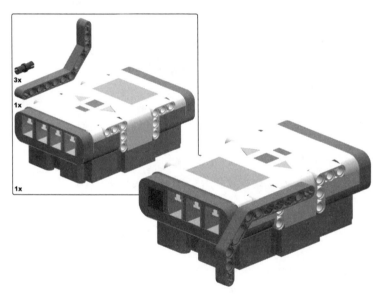

STEP 1 Add parts as shown.

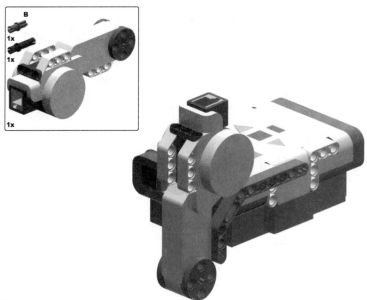

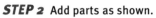

STEP 2 Add parts as shown.

STEP 3 Add parts as shown.

STEP 4 Add parts as shown.

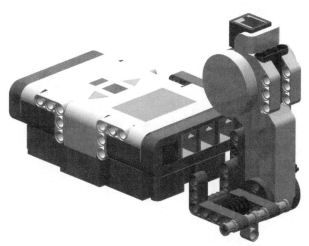

STEP 5 Once you place the worm-gear on the axle in this step, push the axle about 1/2 unit past the worm-gear so it is partially in the hole of the L-beam.

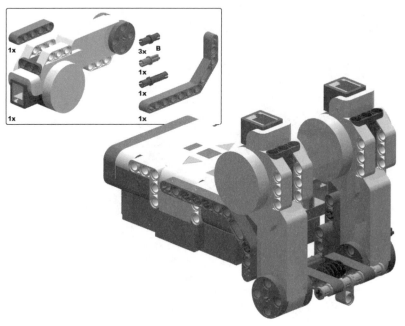

STEP 6 Attach the motor the same as the other side.

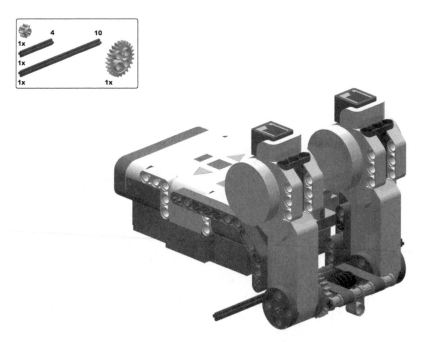

STEP 7 The 4-unit axle goes into the motor and attaches the 24-tooth gear. The 10-unit axle goes through the 5-beam, then the small gear meshes with the 24-tooth gear (hidden) then it goes into the axle-extender, creating one long axle.

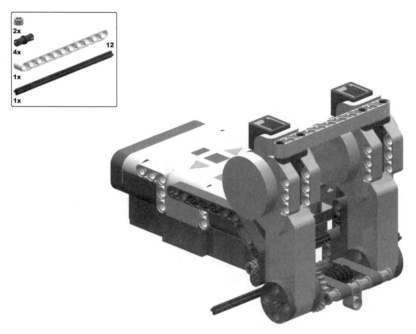

STEP 8 The 12-unit axle goes through everything, with half-bushes on the ends.

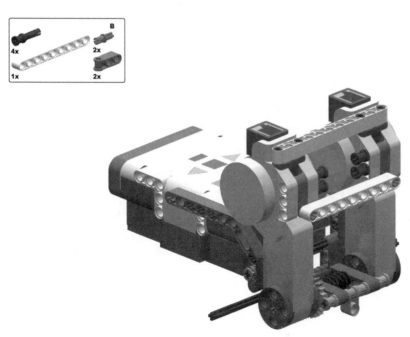

STEP 9 Add parts as shown.

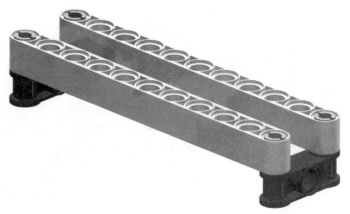

STEP 10 Add parts as shown.

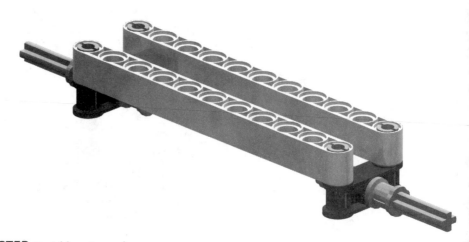

STEP 11 Add parts as shown.

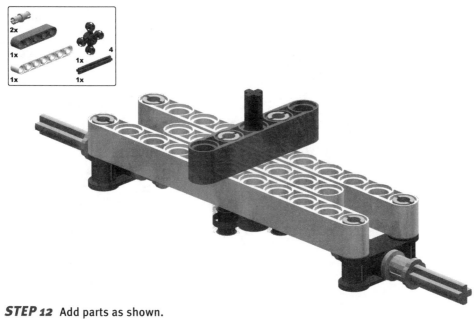

STEP 12 Add parts as shown.

STEP 13 Add parts as shown.

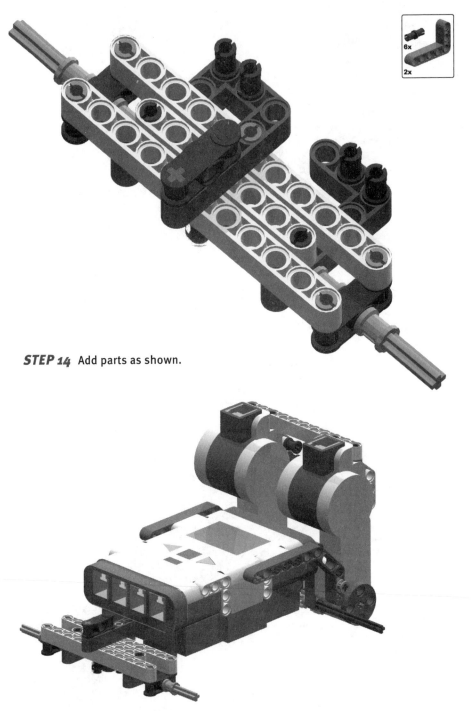

STEP 14 Add parts as shown.

STEP 15 Attach the steering linkage to the underside of the NXT brick.

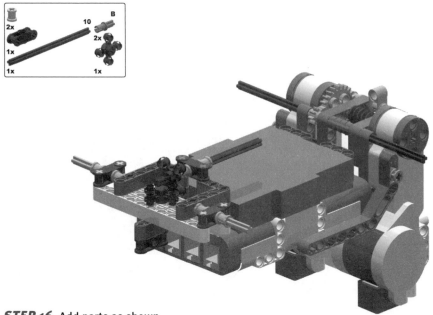

STEP 16 Add parts as shown.

STEP 17 Add parts as shown.

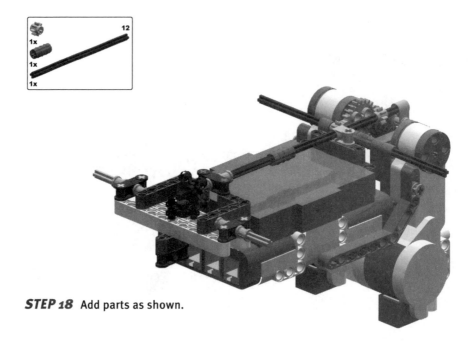

STEP 18 Add parts as shown.

STEP 19 Add the wheels and you're done!

Programming the Moon Buggy

This code allows a user to use the keyboard to drive the moon buggy around. The code for the Moon Buggy is quite simple. It uses the java. awt.event package to listen for key presses. A small frame is generated so the Java code has somewhere to receive key events.

```
1.  import icommand.nxt.*;
2.  import icommand.nxt.comm.NXTCommand;
3.  import java.awt.event.*;
4.  import java.awt.*;
5.
6.  public class MoonBuggy extends Frame implements
        KeyListener{
7.
8.      final int FORWARD = 87, // W = forward
9.      BACKWARD = 83, // S = backward
10.     LEFT = 65, // A = left turn
11.     RIGHT = 68, // D = right
12.     QUIT = 81; // Q = quit
13.
14.     public MoonBuggy(String title) {
15.         super(title);
16.         NXTCommand.open();
17.         Motor.B.setSpeed(900); // Steering
18.         Motor.C.setSpeed(900); // Drive motor
19.         this.setBounds(0, 0, 300, 50);
20.         this.addKeyListener(this);
21.         this.setVisible(true);
22.     }
23.
24.     public void keyPressed(KeyEvent e) {
25.         switch(e.getKeyCode()) {
26.         case FORWARD:
27.             Motor.C.backward();
28.             break;
29.         case BACKWARD:
30.             Motor.C.forward();
31.             break;
32.         case LEFT:
33.             Motor.B.forward();
34.             break;
35.         case RIGHT:
36.             Motor.B.backward();
37.             break;
38.         }
39.     }
40.
41.     public void keyReleased(KeyEvent e) {
42.         switch(e.getKeyCode()) {
43.             case FORWARD:
```

```
44.          case BACKWARD:
45.              Motor.C.flt();
46.              break;
47.          case LEFT:
48.          case RIGHT:
49.              Motor.B.stop();
50.              System.out.println("Tacho: " + Motor.
                   B.getTacho());
51.              break;
52.          case QUIT:
53.              NXTCommand.close();
54.              System.exit(0);
55.          }
56.      }
57.
58.      public void keyTyped(KeyEvent e) {}
59.
60.
61.      public static void main(String[] args) {
62.          new MoonBuggy("Enter commands");
63.      }
64. }
```

Results

Turn on the NXT brick and then run the iCommand code from your
computer. Use the W and S keys to drive forward and backward. Use A
and D to turn left and right. When you are done, press Q. As previously
mentioned, the steering radius is not sharp. Try not to over-steer to make
up for this or the gears will start exerting some adverse forces on the model.

WARNING: *When the battery level starts to get low, Bluetooth communications become unreliable. Make sure the batteries are charged.*

TRY IT: *Using an analog controller makes this robot even better.
You could add pincers to the design and program the joystick buttons
to control it. Sun does not have an official game controller API for
Java, but there are a number of open source solutions available. The
best platform independent solution is JInput: https://jinput.dev.java.net*

TRY IT: *In the popular 80s television show Knight Rider, star David
Hasselhoff drove around in an autonomous vehicle named KITT.
Try adding some sensors to Moon Buggy and changing the existing
code to make the Moon Buggy an autonomous robot.*

In Chapter 16 we will explore more of Bluetooth communications
using the Java Networking API.

Grabby Robots

Topics in this Chapter

- Arm Theory
- Deep Orange – a robot that plays chess
- Futura – an arm that moves in 3D space

Chapter 8

Robot arms are used mostly in manufacturing and have been the most common area of industrial robotics. For example, the car in your driveway was likely partially built by a robot arm. This chapter will explore the concepts needed to successfully build and program arms that can accurately navigate 3D space.

Basic Arm Theory

The human arm is a relatively simple piece of machinery compared to the human hand. The arm is like a big finger, with a shoulder joint that can rotate along two axes, and an elbow that can rotate along one axis. Like other parts of your body, the arm uses proprioception. In other words, you can sense the position of your arm even when you aren't looking at it.

TRY IT: Try paying attention to your own sense of proprioception. First, place a non-breakable object on the table in front of you and hang your hand down by your side. Look at the object, close your eyes and then try to pick up the object. Chances are you were able to pick it up on the first try. Your body tells your brain the approximate angle of each joint in your arm, allowing you to maneuver your arm into position without looking. Even though the robot arms described in this chapter are blind, they can accurately move to 3D coordinates just like your human arm.

There are at least two joints in a robot arm, the shoulder and the elbow. The shoulder is sometimes capable of rotation in two axes, which requires two dedicated motors. As a result, robot arms typically require at least three motors, not including wrist or claw movements.

The controlling program must keep track of the angle of every joint in a robot arm in order to calculate movements in 3D space. The angles are calculated using trigonometry, a high-school level discipline that people tend to forget if they haven't used it in a while.

NOTE: *Appendix B explains the basic trigonometry calculations. If you aren't already familiar with trigonometry it should be considered required reading for robotics programmers.*

This chapter will guide you through the construction of two robots using only the parts in the NXT kit. The first robot plays chess, a tall order for any robot, much less one built from LEGO. The second robot arm described is more generalized, and is capable of manipulating objects in 3D space.

Chess Playing Robot

The robot in this section is called Deep Orange, in homage to Deep Blue, an IBM computer capable of beating chess masters. Even though Deep Orange is built from LEGO, it is a large-scale robot. The arm is about as long as my own arm, and capable of reaching all the way past the edge of a large chessboard.

Although a complete chess program could fit in the NXT memory*, this project will not show how to program a game of chess, which is beyond the scope of this book. Instead, you will have to rely on an open source chess program to control player moves. Our program will merely instruct the arm where to move, and when to open and close the claw.

Building Deep Orange

Robot arms are both easily built and accurately controlled with LEGO NXT. The servo motors allow the NXT to precisely control and keep track of the position of the robot arm. The chess arm has the following parts: base, main arm, forearm, and claw (see Figure 8-1). Each of these parts is designed as a modular component.

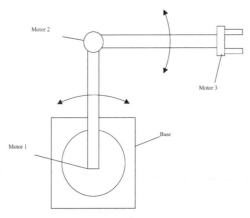

Figure 8-1 An abstract view of a chess arm

---

* The smallest game of chess ever programmed was Microchess, made by Peter Jennings. He programmed Microchess for an early microcomputer called the KIM-1. He was somehow able to program it with only 1118 bytes of memory (not kilobytes).

Let's review the list of requirements for Deep Orange:

- The claw must be able to pick up a chess piece without interfering with the surrounding pieces.
- The claw must be able to pick up pieces varying in height from 3 cm to 7 cm, with different shapes and random orientation.
- The arm must be able to move over the entire surface of a chess board without interfering with other pieces, even while holding a piece.
- The claw must be able to drop a piece without interfering with the surrounding pieces.
- The arm must only use three NXT motors.

It's time to build the arm. Although this model is complicated, it's simple to build because you are adding only a few parts per step.

STEP 1 Add parts as shown.

STEP 2 Add parts as shown.

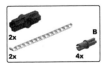

STEP 3 Add parts as shown.

STEP 4 The L-beam on the far side attaches with 4 pins. The L-beam on this side uses only three (the long beam on this side hangs free for now).

STEP 5 The axle now supports the long beam.

STEP 6 Attach the L-beam using one blue axle-pin. It hangs free for now.

STEP 7 Turn the assemblage around and add the 3-unit beam, as shown.

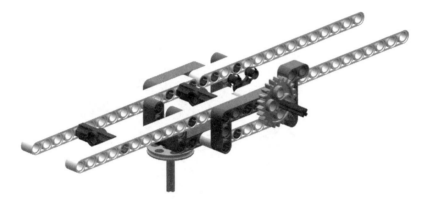

STEP 8 The axle is inserted into the black axle-hole and supported by a bush (hidden).

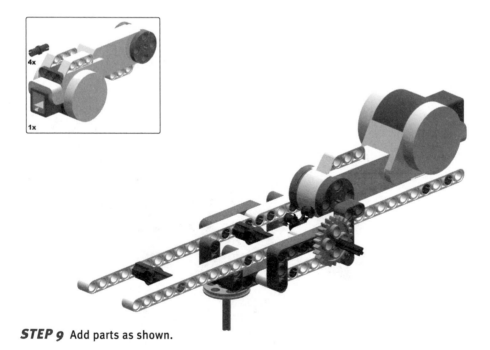

STEP 9 Add parts as shown.

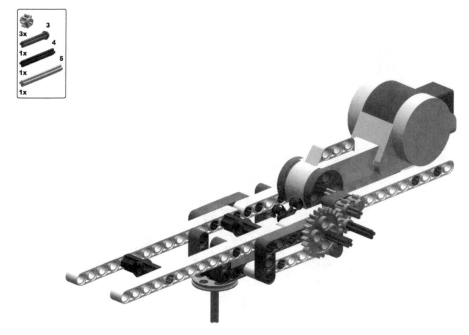

STEP 10 Insert the light-grey 5-unit axle into the motor, then attach a gear. Insert the 3-unit axle with stud through the top hole and place a gear on the axle. The 4-unit axle sits freely for now, with the remaining gear.

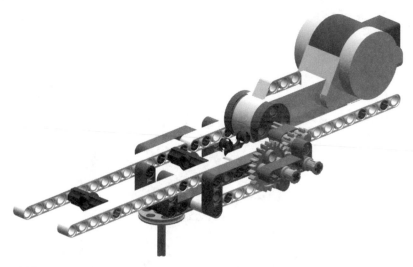

STEP 11 Add parts as shown.

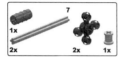

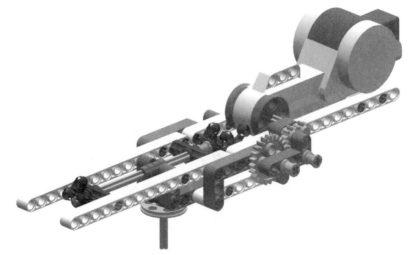

STEP 12 Add parts as shown.

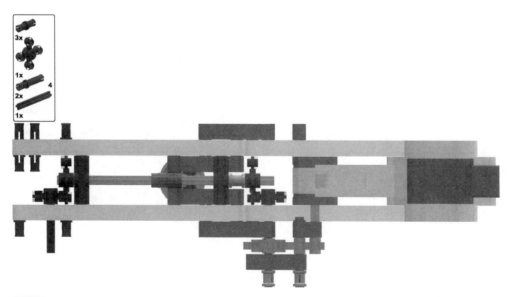

STEP 13 Add parts as shown.

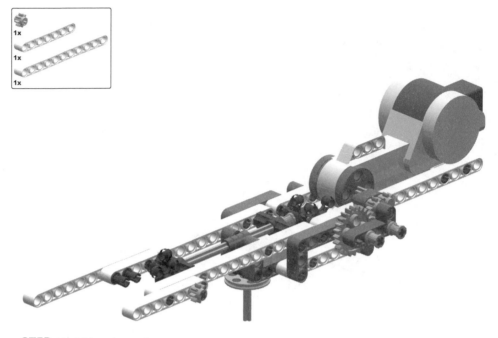

STEP 14 Add parts as shown.

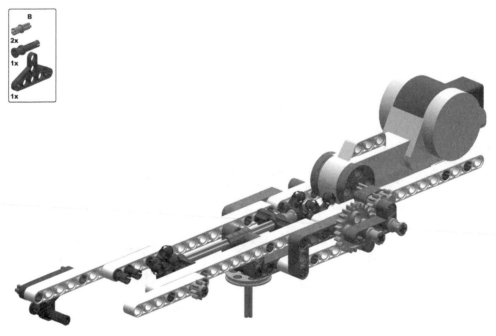

STEP 15 Add parts as shown.

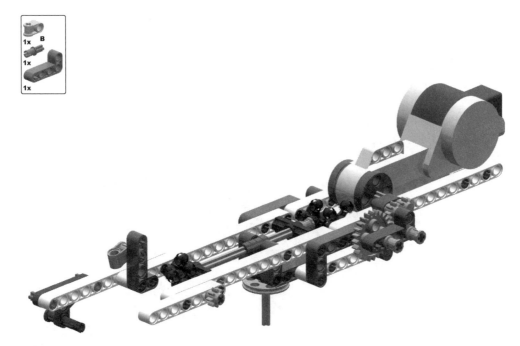

STEP 16 Add parts as shown.

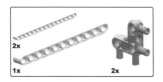

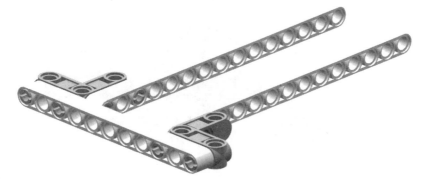

STEP 17 Add parts as shown.

STEP 18 Add parts as shown.

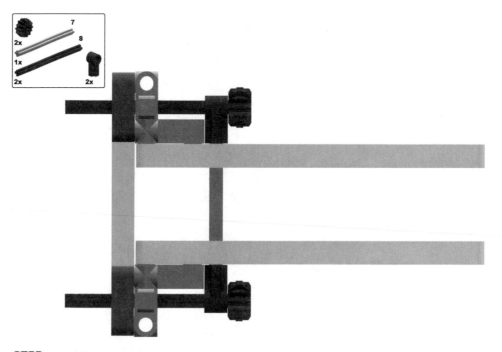

STEP 19 Add parts as shown.

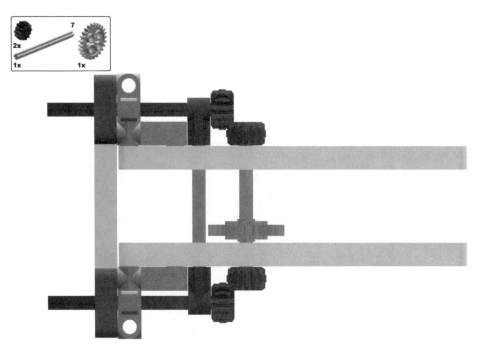

STEP 20 Add parts as shown.

STEP 21 Add parts as shown.

STEP 22 Add parts as shown.

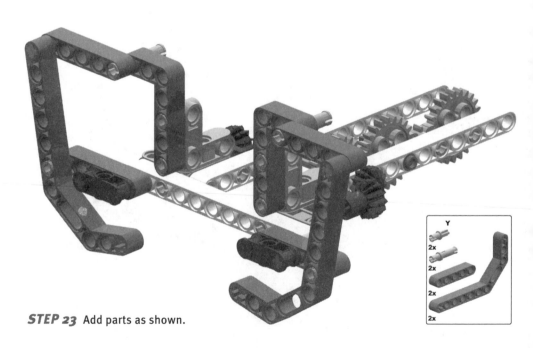

STEP 23 Add parts as shown.

STEP 24 Add parts as shown.

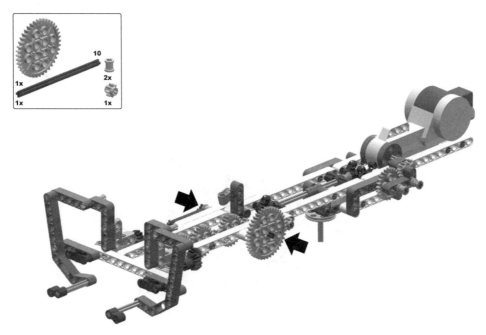

STEP 25 The 10 unit axle has the large gear on the end. The small gear meshes with the 24-tooth gear from step 21. The two bushes are also placed on this axle— one on either side of the beams opposite the large gear (partially obscured).

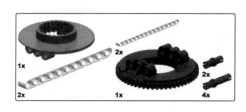

STEP 26 Add parts as shown.

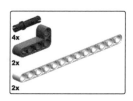

STEP 27 Add parts as shown.

STEP 28 Add parts as shown.

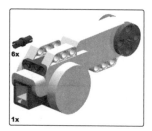

STEP 29 The motor attaches with only two pins as shown. Attach the other 4 pins to the ends of the beams as shown.

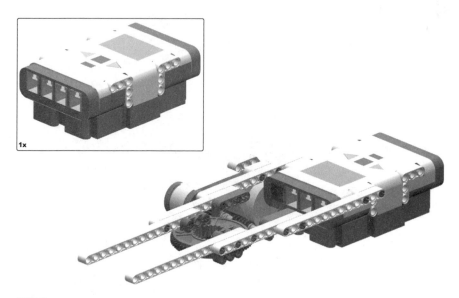

STEP 30 Add parts as shown.

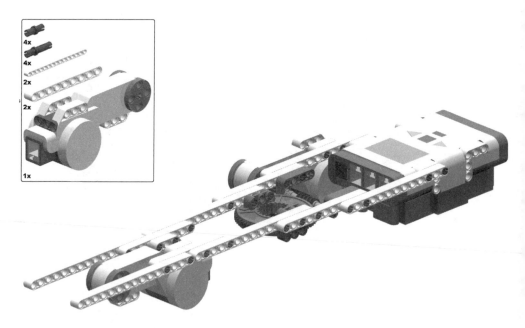

STEP 31 Add parts as shown.

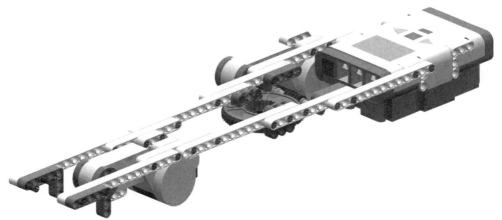

STEP 32 The L-beam closest uses short pins. The farthest L-beam uses long pins.

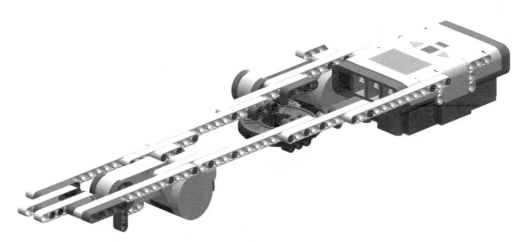

STEP 33 Insert one 3-unit beam between the two white beams using long pins. Add the black part as shown using two 3-unit axles. Add the other 3-unit beam to this side of the assemblage using long pins. Now insert the assemblage onto the main arm.

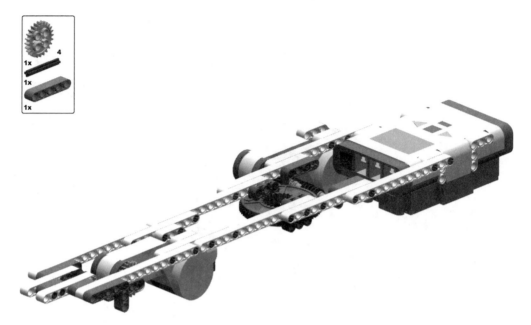

STEP 34 Insert the axle into the motor and attach the gear. The 5-unit beam rest in place for now.

STEP 35 This is a front view to show the parts more clearly along the axle.

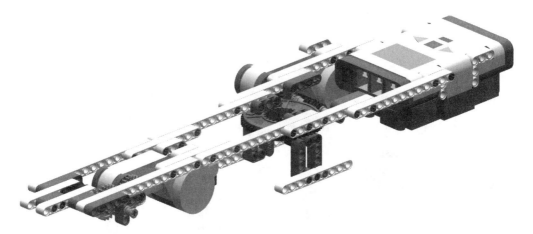

STEP 36 Repeat the same for the other side.

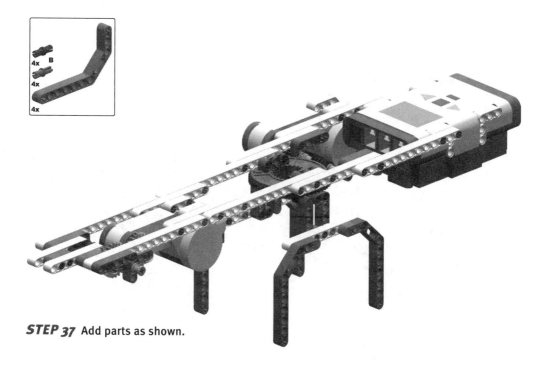

STEP 37 Add parts as shown.

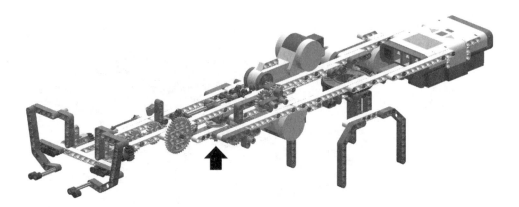

STEP 38 Insert the axle into the black hole of the main arm, then place a gear on the bottom to secure it. (The gear meshes with the worm-screw.)

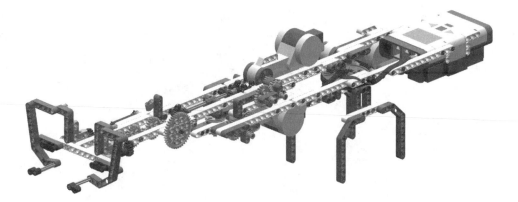

STEP 39 The short cable runs from port A to the lower motor. Use a long pin with bush to keep it off to the side.

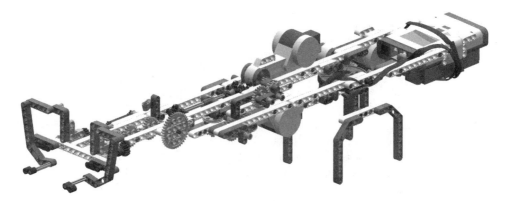

STEP 40 A long cable goes from port B, wraps under the brick, then to the side motor.

STEP 41 A long cable goes from port C to the top motor, hanging freely.

Deep Orange performs a balancing act, with the NXT brick and a motor acting as the main counterweight to the arm, and the other motor acting as a counterweight to the claw.

The rotation of the shoulder and elbow are similar. The shoulder joint uses a simple worm screw to slowly rotate the turntable, causing the whole arm to rotate like a crane. Likewise, a worm screw turns the gear that attaches to the forearm.

The claw itself has two axles, one for each side of the claw. Bevel gears transfer the torque of the main axle to two axles. This also ensures that the fingers on the claw are synchronized.

The most complex part of the arm is the lift mechanism (Figure 8-2). Ideally we would use two motors, one to open and close the claw, and one to lift the chess piece. Since we only have one motor, the claw must grab and lift the piece with one shared motor. When the claw closes on a piece (or on itself) the bevel gears can no longer turn. Normally, this would cause the motor to slow down and the gears to start slipping.

But notice the axle that allows the wrist to bend upwards (see Figure 8-2A). When the claw is closed on a chess piece, the gear train becomes locked up to the wrist joint, while the axle can still rotate against the big gear. This means the axle is the only part that can now move, and it moves upwards until it hits the small stopper piece (see Figure 8-2B). With this design, you get two axle rotations for the price of one.

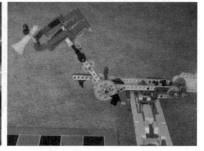

Figure 8-2A The claw clamps down on a chess piece

Figure 8-2B The claw lifts the chess piece

Programming Deep Orange

Our main programming goal is to command the chess arm to move the claw to specific x and y coordinates; we give the program a coordinate (two numbers) and it responds by moving the claw to that coordinate. The position of the claw is determined by two things: the angle of the shoulder and the angle of the elbow. In other words, we give it two numbers (x and y coordinates), and it uses these numbers to calculate two new numbers (shoulder angle and elbow angle). In order to do this, we will use the triangle math described in Appendix A.

We start by knowing x and y. However, we don't know the angles we need to rotate the motors to in order to make the arm move to these coordinates. It helps to visualize where the robot arm should be (see Figure 8-3). In this view we can clearly see that the arm makes up a triangle consisting of three points: motor 1, motor 2, and the chess piece/claw. This is why we use triangle math to determine the angles for the two motors.

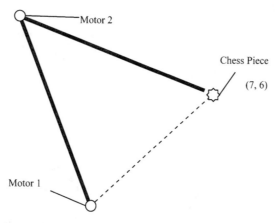

Figure 8-3 Imagining a triangle from the arm

Each side of the triangle has a length. Two lengths are known (the length of the forearm and main arm) and the final length is unknown. We will label these a, b, and c (see Figure 8-4). The angle between a and b is labeled A. As explained in Appendix A, the *law of cosines* has the following equation as it applies to triangles:

$$c^2 = a^2 + b^2 - 2ab \cos A$$

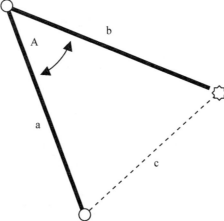

Figure 8-4 Labeled sides of the triangle

If we could just work out the distance of c we would know every variable in this equation except for A. So how do we find c? Since we know the values of x and y (for example, let's use x=7, y=6 as coordinates), we can create a right-angled triangle that includes c (see Figure 8-5). Now that we have this triangle, we can calculate c using:

$$c^2 = a^2 + b^2$$

Rewritten as...

$$c = \sqrt{a^2 + b^2}$$

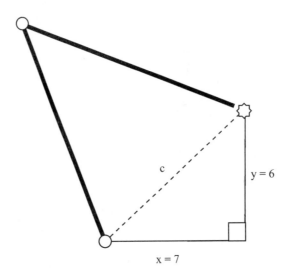

Figure 8-5 Creating a right-angled triangle

Now that we know a, b, and c we can revisit our first equation. We want to calculate A, so we rewrite the equation as follows:

$$A = \text{acos}\left(\frac{a^2 + b^2 - c^2}{2ab}\right)$$

I won't bore you with the equations for the base angle where the main arm rotates, since it is similar to the other angle calculation. All you need to know is that the base angle is determined by adding B1 and B2 (see Figure 8-6), since the arm will start at angle 0, which points East in a Cartesian coordinate system.

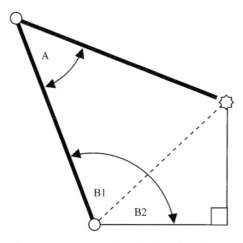

Figure 8-6 Determining the base angle

Now that we know how to calculate the angles, we need to give the program several measurements so it can perform the calculations. We must be precise to achieve accuracy. These measurements include:

- The length of the main arm from the shoulder axis to the elbow axis. (21.0 cm)
- The length of the forearm from the elbow axis to the claw. (24.5 cm)
- The distance from the center of the shoulder axis to the edge of the chess board. (4.6 cm)
- The border around the chessboard. (Depends on your chessboard)
- The length of board from one side to the other, not including the border. (Depends on your chessboard)

We will program Deep Orange using iCommand, since you may want to try to interface this code with a real thinking chess game. The code to perform these calculations is as follows:

```
1. import icommand.nxt.*;
2.
3. public class ChessRobot {
4.
5.     // Tachocount for full 360 degree = 56:1 * 360 = 20160
6.     static final int SHOULDER_FULL_CIRCLE = 56 * 360;
7.     // Tachocount for full 360 degree rotation = 24:1 *
           360 = 8640
8.     static final int ELBOW_FULL_CIRCLE = 24 * 360;
9.
10.    static int NEUTRAL_CLAW = 0;
11.    static int LIFT_CLAW = 3800;
12.    static int OPEN_CLAW = -5000;
13.
```

```
14.     static double ARM_BASE_LENGTH = 21; // length of base
           arm (shoulder to elbow) cm
15.     static double ARM_FORE_LENGTH = 24.5; // length of
           fore arm (elbow to claw) cm
16.
17.   static double BOARD_BORDER = 2.5; // Distance from
           outer edge of board to inner edge of square
18.     static double BOARD_WIDTH = 35.7;
19.     static double SQUARE_SIZE = BOARD_WIDTH / 8; //
           Width/length of square
20.     static double BOARD_TO_AXIS = 4.6; // Distance from
           shoulder axis to edge of board
21.
22.     public static void main(String[] args) {
23.         icommand.nxt.comm.NXTCommand.open();
24.
25.         ChessRobot c = new ChessRobot();
26.         c.movePiece(2,5,4,5);
27.         c.movePiece(8,7,6,6);
28.
29.         Motor.B.rotateTo(0); // Reset shoulder
30.         Motor.A.rotateTo(0); // Reset elbow
31.         Motor.C.rotateTo(0); // Reset claw
32.
33.         icommand.nxt.comm.NXTCommand.close();
34.     }
35.     /** Columns 0-7 and rows 0-7? */
36.     public void movePiece(int fromRow, int fromColumn,
           int toRow, int toColumn) {
37.         openClaw();
38.         gotoSquare(fromRow, fromColumn);
39.         grabPiece();
40.         gotoSquare(toRow, toColumn);
41.         openClaw();
42.     }
43.
44.     public void gotoSquare(int row, int column) {
45.         double x = (row * SQUARE_SIZE) - (BOARD_WIDTH/2)
               - (SQUARE_SIZE/2);
46.         double y = BOARD_BORDER + BOARD_TO_AXIS + BOARD_
               WIDTH - (column * SQUARE_SIZE) + (SQUARE_SIZE/2);
47.         gotoXY(x, y);
48.     }
49.
50.     public void openClaw() {
51.         Motor.C.rotateTo(OPEN_CLAW);
52.     }
53.
54.     public void grabPiece() {
55.         Motor.C.rotateTo(LIFT_CLAW);
```

```
56.      }
57.
58.      public void gotoXY(double x, double y) {
59.          double c = Math.sqrt(x * x + y * y);
60.          double angle1a = Math.asin(y/c);
61.          double angle1b = Math.acos((ARM_BASE_LENGTH *
               ARM_BASE_LENGTH + c * c - ARM_FORE_LENGTH *
                 ARM_FORE_LENGTH)/(2 *ARM_BASE_LENGTH * c));
62.          double angle1 = angle1a + angle1b;
63.          double angle2 = Math.acos((ARM_BASE_LENGTH * ARM_
               BASE_LENGTH + ARM_FORE_LENGTH * ARM_FORE_LENGTH
                 - c * c)/(2 *ARM_BASE_LENGTH * ARM_FORE_LENGTH));
64.
65.          if(x < 0) {
66.              angle1 = Math.PI - angle1;
67.              angle2 = Math.PI - angle2 + Math.PI; // Untested!!
68.          }
69.
70.          rotateShoulder(angle1);
71.          rotateElbow(angle2);
72.      }
73.
74.      public void rotateElbow(double toAngle) {
75.          // Arm starts at tacho 0 which is actually 180
                 degrees, hence subtract 1/2 ELBOW_FULL_CIRCLE
76.          int toCount = (int)((toAngle/(2*Math.PI)) * ELBOW_
               FULL_CIRCLE) - ELBOW_FULL_CIRCLE/2;
77.          Motor.A.rotateTo(toCount);
78.      }
79.
80.      public void rotateShoulder(double toAngle) {
81.          int toCount = (int)((toAngle/(2*Math.PI)) *
               SHOULDER_FULL_CIRCLE);
82.          Motor.B.rotateTo(toCount);
83.      }
84. }
```

Results

Before we can test Deep Orange we must position the arm relative to the board. First, place the board with the white pieces along the bottom (the side closest to you) and the black pieces along the top. With the white pieces closest to you, position the arm on the right side of the board with the arm pointing away from you (see Figure 8-7).

The arm must be perfectly straight, with the claw closed and hanging at the neutral position. Make sure the base of the arm is touching the board. Now measure the width of the border, and the width of the chessboard playing area. Input these into the program at lines 17 and 18.

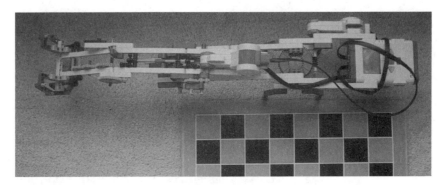

Figure 8-7 Positioning the arm relative to the board

 TIP: *If your chess pieces are slippery or the claw is having trouble gripping them, try placing the rubber axle connectors all along the claw axles for improved grip.*

You are now ready to try out the arm. Place one chess piece on square 8,7 and another on square 6,6, as shown in Figure 8-8. Now run the program.

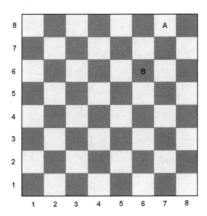

Figure 8-8 Placing the chess pieces

The forearm is a balancing arm and the cable throws the balance off, sometimes causing the arm to go too high or too low when grabbing a piece. This was an unexpected dose of reality on this project.

The cable is just long enough to reach the farthest motor, but it is so stiff that it affected movement. I found I needed to manually manage the cable, making sure it kept out of the way and didn't hinder rotation.

TIP: *If you are serious about perfecting your chess arm, you can obtain ribbon cables from Mindsensors.com. The ribbon cable is less rigid, and interferes less with the arm movement. See Appendix A for more information.*

 TRY IT: *Using an open source Java chess program, hack the code and attempt to intercept calls to the chessboard so that your arm physically moves the pieces. Sourceforge is a good place to look for open source chess programs.*

 TRY IT: *Make the chess-playing robot into a building robot that can stack up small blocks into structures. Ideally, you will feed it a construction plan and it will build the structure.*

Arms with Three Degrees of Movement

The chess arm was really a robot arm in 2D space, capable of movement along the x- and y-axes, but not really the z-axis since it couldn't lift the claw a significant or predetermined amount. The arm in this section uses three motors to move in the claw along all three axes of x, y, and z.

Building the Arm

The arm in this section is named Futura, after the robot in the film *Metropolis* (1927). Futura is extremely simple to build and requires few parts, mostly because of the excellent NXT servo motors. Let's go over some goals for this project.

- Fast—able to perform movements as fast as a human
- Powerful
- Far Reaching
- Strong—parts won't break off or fall apart
- Reliable—cord won't tangle up

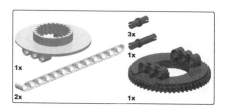

STEP 1 This side uses a long and short pin to hold the beam. The other side uses two short pins.

STEP 2 Rotate the assemblage. Place the two grey L-beams as shown, then add the 5-unit beams.

STEP 3 Rotate the assemblage. Add the parts as shown.

STEP 4 Rotate the assemblage. Insert the axle into the center hole, with the knob-gear on the bottom (obscured) and the small gear on top. Add the angle-connector using a blue pin, as shown.

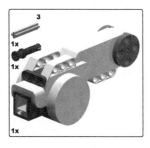

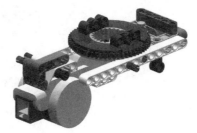

STEP 5 The motor attaches to the small L-beam.

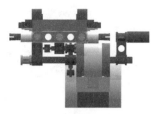

STEP 6 Insert the axle through the motor, through a knob-wheel, through the angle-connector and secure with a bush as shown. A small hand crank is added using the remaining parts, as shown.

STEP 7 Add four bent-beams as shown. The two far beams attach using a blue axle-pin and a black pin. The closer beams are attached as shown.

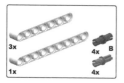

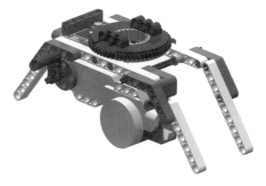

STEP 8 Add parts as shown.

STEP 9 Add parts as shown.

STEP 10 Rotate the assemblage.

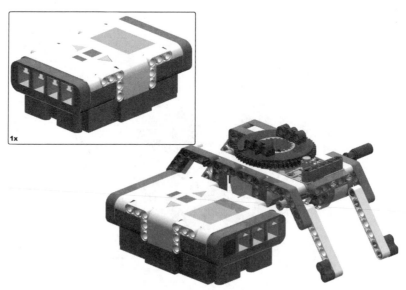

STEP 11 Add parts as shown.

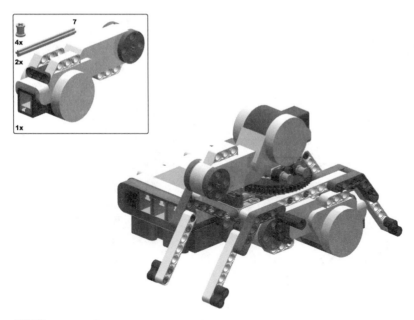

STEP 12 Attach the motor using axles and bushes.

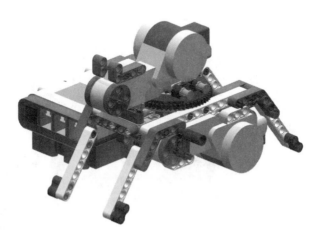

STEP 13 Add parts as shown.

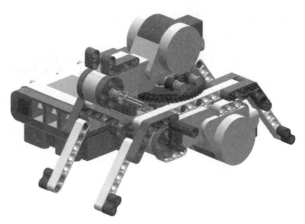

STEP 14 Add parts as shown.

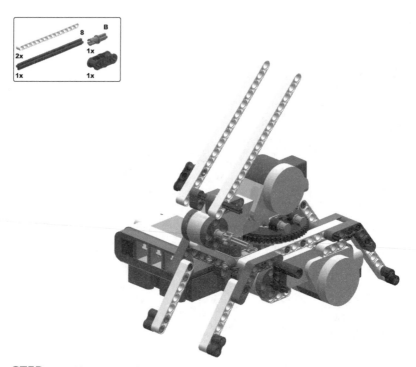

STEP 15 Add parts as shown.

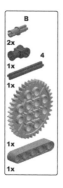

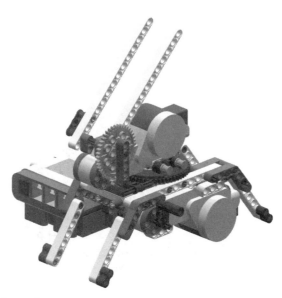

STEP 16 Add parts as shown.

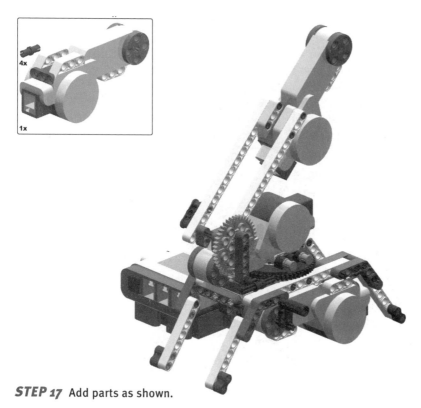

STEP 17 Add parts as shown.

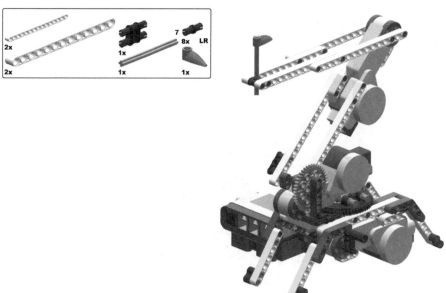

STEP 18 All done!

Programming Futura

The triangle math must determine three angles as opposed to the two angles of the chess arm. Figure 8-9 shows the top view of the arm. With this view, only the z-axis will be seen to move because the elbow will always appear straight. Figure 8-10 shows the side view. From this angle, we can see two joints: A2, the shoulder joint, and A3, the elbow joint. The complete 3D coordinates are calculated using all three of these angles. Because we discussed triangle math so extensively in the earlier section we can skip this explanation and go straight to code. The code for this project uses leJOS NXJ.

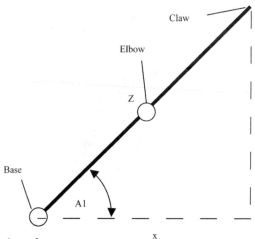

Figure 8-9 Top view of arm

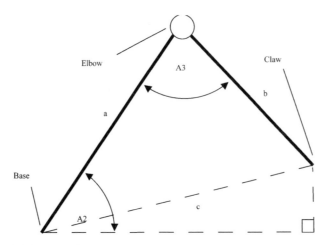

Figure 8-10 Side view of arm

```
1.  import lejos.nxt.*;
2.
3.  public class Futura {
4.
5.      double BASE_ARM_LENGTH = 16.3; // cm's
6.      double FOREARM_LENGTH = 16.6;
7.      double DISTANCE_FROM_CENTER = 6.5; // dist from
            center axis
8.
9.      int ZAXIS_START_ANGLE = 0;
10.     int SHOULDER_START_ANGLE = 135;
11.     int ELBOW_START_ANGLE = 20;
12.
13.     public static void main(String [] args) throws
          Exception{
14.        Futura arm = new Futura();
15.        arm.gotoPoint(10, 30, 5);
16.        arm.gotoPoint(30, 1, 15);
17.        arm.gotoPoint(20,25, 0);
18.
19.        Motor.A.rotateTo(0);
20.        Motor.B.rotateTo(0);
21.        Motor.C.rotateTo(0);
22.     }
23.
24.     public Arm() {
25.        Motor.C.setSpeed(150);
26.        Motor.B.setSpeed(500);
27.        Motor.A.setSpeed(600);
28.     }
29.
30.     public void gotoPoint(double x, double y, double z) {
```

```
31.          // 1. Figure out Z-Axis angle on shoulder
32.          double zaxisangle = Math.atan2(y,x);
33.
34.          // 2. Figure out shoulder angle:
35.          double Z = Math.sqrt(y * y + x * x);
36.          Z = Z - DISTANCE_FROM_CENTER; // Corrected due to
                 arm construction
37.          double c = Math.sqrt(Z * Z + z * z);
38.          double angle1 = Math.asin(z/c);
39.          double angle2 = Math.acos((BASE_ARM_LENGTH
               * BASE_ARM_LENGTH + c * c - FOREARM_LENGTH *
                 FOREARM_LENGTH)/(2*BASE_ARM_LENGTH*c));
40.          double shoulderangle = angle1 + angle2;
41.
42.          // 3. Figure out elbow angle:
43.          double elbowangle = Math.acos((Math.pow(BASE_ARM_
               LENGTH, 2) + Math.pow(FOREARM_LENGTH, 2) - Math.
                 pow(c, 2))/(2*BASE_ARM_LENGTH*FOREARM_LENGTH));
44.
45.          rotateZAxisTo(Math.toDegrees(zaxisangle));
46.          rotateShoulderTo(Math.toDegrees(shoulderangle));
47.          rotateElbowTo(Math.toDegrees(elbowangle));
48.      }
49.
50.      /**
51.       * Rotate elbow up or down.
52.       * @param angle +ve value is up.
53.       */
54.      public void rotateElbowTo(double angle) {
55.          angle = angle - ELBOW_START_ANGLE;
56.          if(angle < 0)
57.             return; // Angle too small
58.          Motor.C.rotateTo((int)-angle);
59.      }
60.
61.      /**
62.       * Rotate shoulder up or down.
63.       * @param angle +ve value is up.
64.       */
65.      public void rotateShoulderTo(double angle) {
66.          angle = angle - SHOULDER_START_ANGLE;
67.          Motor.B.rotateTo((int)angle * 5); // Gear ratio 40:8
68.      }
69.
70.      /**
71.       * Rotate base of arm.
72.       * @param angle +ve value is counterclockwise
73.       */
74.      public void rotateZAxisTo(double angle) {
75.          int ratio = 56/8;
```

```
76.        angle = angle - ZAXIS_START_ANGLE;
77.        Motor.A.rotateTo((int)(-angle * ratio));
78.    }
79. }
```

Results

Use the handles on the model to move the arm into the neutral position. This looks something like a cobra striking. You can now run the program and the arm will quickly move to the designated coordinates.

NOTE: *Make sure to set the arm to the starting position before you turn on the NXT. If you turn it on and then set the starting position, the motors will keep track of the tachometer changes and will throw the code off.*

I really like the speed and purpose with which Futura moves. It reminds be of industrial robot arms that are used to weld cars (though probably a little less precise). You can use any coordinates, even negative coordinates that move outside of quadrant I.

Let's try giving Futura an actual task. Build several rings as shown in Figure 8-11. Set these in the test area at different coordinates and see if you can make the hook pick them up.

Figure 8-11 LEGO rings

TRY IT: *Try attaching a magnet to the end of the arm instead of a hook and see if you can instruct the robot to pick up metallic objects.*

Sound

Topics in this Chapter

- Playing a tone
- Playing prerecorded sound
- Playing tunes
- Recording sound

Chapter 9

S ound is not vital to robotics programming, but it can aid with robot to human communication. You won't be having deep conversations with a robot by the end of this chapter, however sound can tell you what the robot is sensing. For example, when the robot detects movement, it can say a meaningful phrase such as "I see you".

This chapter will study the different ways you can use sound with leJOS NXJ. Some are useful, some are just for fun.

Playing a Tone

The NXT can generate a simple tonal sound. In leJOS NXJ there are several methods in the lejos.nxt Sound class to play different sounds. Sound.playTone() will play a tone by accepting an argument for frequency and duration. The following code plays an ascending series of notes.

```
 1. import lejos.nxt.*;
 2. public class SoundTest {
 3.
 4.   public static void main(String[] args) throws Exception {
 5.     for(int i = 3; i<20; i++) {
 6.     Sound.playTone(i*100,1000);
 7.     Thread.sleep(1000);
 8.     }
 9.     Button.ESCAPE.waitForPressAndRelease();
10.   }
11. }
```

Recording and Playing

NXT can store and play sound files, with either the standard LEGO firmware or iCommand. It uses a proprietary sound format which uses files with the .*RSO* extension. These sound files use less memory than MP3 and WAV file formats but they also contain less sound information. That is why they don't sound as clear.

TIP: *You can find a collection of RSO sound files in the directory where you installed the LEGO software:* \LEGO Software\LEGO MINDSTORMS NXT\engine\Sounds

There are three steps to playing your own files:

- Record a sound file
- Convert the sound file to RSO
- Upload the file to the NXT

Recording a File

Recording a file is simple; you just need to connect a microphone or headset to your computer.

TIP: *To make sure the microphone level is recording properly, go to* Start > Control Panel > Sounds and Audio Devices. *Make sure Sound is not muted, then click the Audio tab. Under Sound recording, click Volume... and make sure the volume is 100% and not muted. Sometimes, when a Bluetooth headset is added to Windows XP, it mutes the device by default, so make sure to check this setting.*

Windows

1. Click on Start > All Programs > Accessories > Entertainment > Sound Recorder (Figure 1).
2. Click the red button and speak into the microphone.
3. Click File > Save and save the WAV file to a project directory.

Figure 9-1 Windows Sound Recorder

Linux

1. A popular sound recording program for Linux is KRec, but it is often not installed by default. Install this on your system and run it from the Multimedia program group.
2. Select File > New and use the default settings for this file, but choose Mono instead of Stereo.
3. When you are ready to record, click the red Record button and speak into the microphone (see Figure 9-2). Hit Stop when done.
4. Select File > Export... and save the file with a .wav extension. Save it anywhere for the time being.

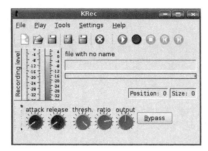

Figure 9-2 KRec under KDE

Macintosh

1. Mac OS X users can use GarageBand if it is installed on the system, or any version of Quicktime Player after 2006.

2. In QuickTime Player, select File, Record Audio...

3. When you are ready to record, click the Record button and speak into the microphone. Hit Stop when done.

4. Save the audio file in .wav format.

Figure 9-3 The WAV2RSO Application (Windows version)

Converting a Sound File

Now that you have a WAV file, it's time to convert it. We will use a handy tool, developed by John Hansen, called Wav2Rso, which is available for multiple platforms.

1. Download Wav2Rso from the utilities section at:
   http://bricxcc.sourceforge.net/utilities.html

2. Unzip and Run the program to bring up the user interface (see Figure 9-3).

3. Browse to your sound file and open it. The file will now appear in the list.

4. Select one of the Resample choices. *Sinc 1* reduces the file size by about six times.

5. Click Convert. You will now see your RSO file in the directory.

Uploading and Playing a Sound File

It will take a little bit of Java code to upload and play the file. The code is self explanatory so I will skip an explanation:

```
1.  import java.io.File;
2.  import icommand.nxt.*;
3.  import icommand.nxt.comm.*;
4.
5.  public class PlaySound {
6.
7.     public static void main(String [] args) {
8.        NXTCommand.open();
9.        File myFile = new File("Foundyou.rso");
10.       FileSystem.upload(myFile);
11.       Sound.playSoundFile(myFile.getName());
12.       NXTCommand.close();
13.    }
14. }
```

That's it. Make sure to put the sound file in the same directory as the code. If all goes well, your NXT brick will play the file you recorded. Keep in mind that if you want to rerecord and upload the same filename you have to delete the old file first.

Playing a Tune

A MIDI file contains musical notes for playing music. Converting a MIDI file to a LEGO compatible sound file is easy, and the steps match much of what we covered in the last section.

WEB LINK: *You can find midi files easily on the Internet. Try:* www.mididb.com

When LEGO created the NXT, the firmware was programmed with the ability to play tune files. In the interval between engineering the brick and its release, they decided to remove official support for tune files. But the code remains. This section will take advantage of this undocumented feature.

1. Find a MIDI file and save it to disk (see link above).

2. Download *Midi Batch* from:
   http://bricxcc.sourceforge.net/utilities.html

3. When you run the program you will be presented with a GUI (see Figure 9-4). In the "Input Directories" field, type the directory with your MIDI file(s), then click Add.

4. In the "Output Directory" field, type the directory to store the RMD files.

5. Now click Execute and the application will convert all the MIDI files into RMD files.

6. Midi Batch produces RMD files, which the LEGO NXT will not recognize. To get the NXT to recognize the file, change the extension to RSO. Now move it to the same directory as your PlaySound code.

7. Now upload the file and play it using the same code from the previous section, only replace the filename with your new file name.

```
File myFile = new File("Monkey.rso");
```

 NOTE: *Since MIDI files can play several sounds simultaneously and the NXT code can play only one note, the song might sound strange for all or part of some tunes.*

Figure 9-4 Midi Batch application (Windows version)

This concludes our examination of sound with LEGO NXT. In Chapter 21 we will explore sound again with speech recognition and speech synthesis.

Robots with Vision

Topics in this Chapter

- Video Imaging
- iCommand Vision API
- Telerobotics
- Beckhambot
- Autobot

Chapter **10**

Vision is one of the most exciting and complex components of robotics. There are two basic uses for vision: as telepresence for yourself, and as a sensor for your robot. This chapter will explore both uses, demonstrating how to use a real-time video feed with robots.

Vision and Robotics

The iCommand package contains classes to deal with live video images. The package was initially designed to mimic the basic functions of the LEGO MINDSTORMS Vision Command software. These functions include detecting color, light and motion within regions of the video display.

The Vision package was created by longtime leJOS developer Lawrie Griffiths. He designed the API to be easily extensible so other programmers could add more visual algorithms to the software. So far this hasn't happened, but you might see more functions appear in the future as the leJOS NXJ platform evolves.

Because of the high memory requirements of visual data, you must connect the video camera to your computer and use the iCommand software to control the robot.

Choosing a Camera

What kind of camera do you need? The LEGO Vision Command camera will work, but so will any other web cam. In fact, just about any USB camera is compatible with the vision software. Ideally you should use a wireless camera so your robot can be fully mobile and not hampered by cords.

Wireless cameras are not widely available or cheap; the X10 (see Figure 10-1) is the most common. This camera is not the smallest on the market since it is designed to be wall mounted, but it is available at a relatively low cost (see Appendix A for ordering information).

Figure 10-1 X10 Wireless camera (dime for scale)

You actually need more than just a camera. The X10 camera needs a video receiver and a special USB adapter for the video signal to connect to your PC (see Appendix A for part list). X10 sells some variations on their basic camera. Some models contain built in microphones. There are also models for low-lighting conditions. If you are using your robots primarily in an indoor environment with few lights, you might want to consider this model. There is also a battery case that allows the X10 to run off 4-AA batteries (see Figure 10-2).

Figure 10-2 Battery pack

WARNING: *In Windows, the X10 software installs and runs something called X10nets.exe. This interferes with the Serial Port Profile of Bluetooth. If Bluetooth stops working after installing the X10 software, check if this service is running.*

1.1 *Click Start > Control Panel > Performance & Maintenance > Administrative Tools > Services. You should see a window with a list of services (see Figure 10-3).*

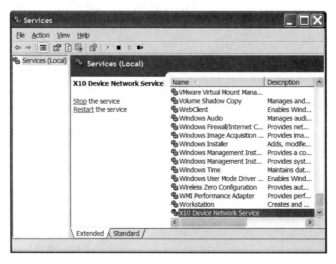

Figure 10-3 X10 in list of services

1.2 *Scroll to the bottom of the list and you might see X10 Device Network Service. If you do, we need to disable it.*

1.3 *Right click the X10 entry and select Properties. You will see more information for the X10 service (see Figure 10-4).*

1.4 *Under Startup Type, change it from Automatic to Disabled. Click Stop to stop the service, and then click OK.*

Figure 10-4 Disabling the X10 service

The X10 camera system uses directional antennas, which means you will not receive even reception in all directions. In practice, as long as the antennas are within a few meters the camera will receive a good picture even if the antennas are facing opposite directions. You can also improve video reception by changing to a different channel if you experience interference.

Although I used an X10 camera, you can use any other wireless camera. Appendix A lists other options for wireless cameras.

NOTE: *You might experience radio interference if you have a lot of wireless devices in your house (see Table 7-1 from the Bluetooth chapter). Table 10-1 shows the four channels used by the X10 camera. Interference can be identified by wavy lines and distortion in the video image, or poor Bluetooth performance. If you experience interference, change to a different channel. For example, every time I moved my Bluetooth mouse the image distorted. I switched the X10 camera from channel A to channel B and the image cleared up (there's a switch on the camera and on the receiver to change channels – both must be changed).*

Channel	Frequency
A	2.411 GHz
B	2.434 GHz
C	2.453 GHz
D	2.473 GHz

Table 10-1 X-10 channel frequencies

Vision API

The Vision API contains more than a dozen classes, but the most important class is called Vision. This class allows you to view live video on your computer screen, take snapshots of the current image, record video, flip the image to suit the orientation on your robot, or add listeners that detect color, light and movement. Let's examine some methods in the Vision class:

icommand.vision.Vision

- `static void addColorListener(int region, ColorListener cl, int color)`

    Add a Color Listener to a region

- `static void addLightListener(int region, LightListener ll)`

    Add a Light Listener to a region

- `static void addMotionListener(int region, MotionListener ml)`
    Add a Motion Listener to a region

- `static void addRectRegion(int region, int x, int y, int width, int height)`
    Add a rectangular region

- `static void flipHorizontal(boolean flip)`
    Flip the image in the image viewer horizontally.

- `static int getAvgBlue(int region)`
    Get the average blue value for the region

- `static int getAvgGreen(int region)`
    Get the average green value for the region

- `static int getAvgRed(int region)`
    Get the average red value for the region

- `static int getAvgRGB(int region)`
    Get the average RGB value for the region

- `static Region[] getRegions()`
    Get an array of regions

- `static boolean isRecording()`
    Check if a recording is in progress

- `static void playSound(java.lang.String fileName)`
    Play an audio file on your computer

- `static void setFrameRate(float rate)`
    Set the video frame rate

- `static void setImageSize(int width, int height)`
    Set the size of the video viewer image

- `static void snapshot(java.lang.String filename)`
    Take a snapshot

- `static void startRecorder(java.lang.String fileName, int millis)`
    Start the video recorder

- `static void startViewer(java.lang.String title)`
    Start the video viewer frame

- `static void stopRecording()`
    Stop the video recorder.

- `static void stopViewer()`

  Close video viewer

- `static void writeImage(java.lang.String fn, byte[] data,`
  `int width, int height)`

  Write the data to the file fn (filename) using the width and height variables.

There are also three Listener interfaces that you can use to react to changes in color, light, or motion. They are:

icommand.vision.ColorListener

- `void colorDetected(int region, int tc)`

  Returns the region number and the color detected. Use Vision. addColorListener to specify the color to detect and the region to scan.

icommand.vision.LightListener

- `void lightDetected(int region)`

  Returns the region number where light was detected. Use Vision.addLightListener to specify the region to scan.

icommand.vision.MotionListener

- `void motionDetected(int region)`

  Returns the region number where motion was detected. Use Vision.addMotionListener to specify the region to scan.

Setting up the Vision API

To use the classes in the icommand.vision package you must first set up the Java Media Framework API (JMF). This package is developed by SUN and helps establish a connection between Java and your video camera.

1. Download the Java Media Framework API:

   http://java.sun.com/products/java-media/jmf/

2. Make sure the camera is plugged into your computer (and turned on if appropriate), then run the Java Media Framework installation. It may take several minutes to detect the camera, even on a fast system (see Figure 10-5). Reboot when told to do so.

Figure 10-5 Discovering video capture devices

3. You are not finished yet. Next you must add a video.properties file to define your camera. Run JMStudio, an application that installed with the JMF. In Windows, click Start > All Programs > Java Media FrameWork > JMStudio.

4. In JMStudio select File > Preferences. Select the Capture Devices tab (see Figure 10-6).

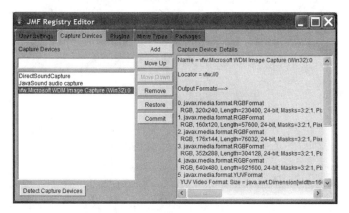

Figure 10-6 Identifying the label for your video camera

5. In the list of Capture Devices, select the image capture device for your camera. In my case, it showed up as:

   ```
   vfw:Microsoft WDM Image Capture (Win32):0
   ```

6. Now open a text editor and enter the name of your preferred sound and video devices (substitute the name you identified in step 5). Also enter the preferred resolution and color-depth of your camera. This can be found in the Capture Device Details information (Figure 10-6). I use 160 x 120 to keep frame rates snappy and because the robot doesn't need detailed visual information. My complete file appears as follows:

   ```
   1. video-device-name=vfw:Microsoft WDM Image Capture
      (Win32):0
   2. sound-device-name=JavaSound audio capture
   3. resolution-x=160
   4. resolution-y=120
   5. colour-depth=24
   ```

NOTE: *Use Direct Sound Capture over Java Sound Capture. The Java sound capture did not work for my system.*

7. Now save the text file in your user directory as video.properties:

    Windows: c:\Documents and Settings\User

    Linux: /home/userid or /users/userid

    MacOSX: /users/userid

8. You can now test your camera with JMStudio. Select File > Capture... and you will see a window with options (Figure 10-7). Make sure your video camera is selected and uncheck sound. You can also change some optional settings, then click ok.

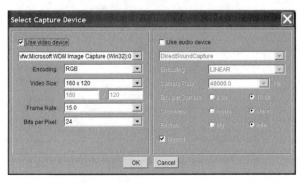

Figure 10-7 Testing the video camera

If everything is working you will see live video from your camera after a few seconds. That's all! You are now ready to use the Vision API.

 NOTE: *Eclipse users will need to add some JMF classpath settings. In Eclipse, select* Project > Properties > Java Build Path. *Click Add External JARs... and add the exact location of the jmf.jar (see Figure 10-8). e.g.* c:\Program Files\JMF2.x\lib\jmf.jar

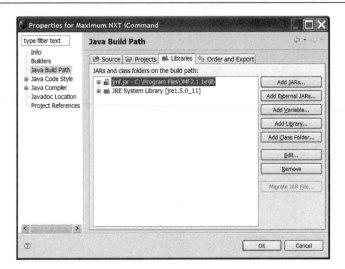

Figure 10-8 Adding the JMF jar to Eclipse

You also need to give Eclipse access to the native DLL's for JMF. In the Java Build Path window, expand jmf.jar and change the native library setting (see Figure 10-9):

`C:\WINDOWS\system32`

If you opted not to copy the native library into this directory when installing JMF, you should browse to the JMF directory.

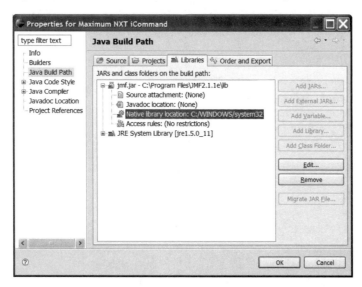

Figure 10-9 Adding the native library location

Telerobotics

This project is a simple example of how a camera allows you to experiment with telerobotics. There are two aspects to telerobotics. The first is teleoperation, which allows you to control a device from a distance. The second is telepresence, which allows a person to feel as if they are at a location other than their actual location.

Telepresence allows you to "be there" without physically being present by giving you detailed visual information. Even if you aren't in the same room, you can control the robot and have it do your bidding. The experience is similar to the Mars Rover missions. NASA used cameras onboard the Sojourner and transmitted images back to earth, allowing earthbound humans to experience vision on Mars without actually being there. Obviously your X10 camera can't broadcast over such long distances, but you should be able to broadcast from different rooms in your home.

Building Beckhambot

Soccer is a simple enough game to adapt to robotics. This robot, named Beckhambot, can move around a play area and kick a LEGO ball. It is not autonomous – yet. Instead, it relies on you to control it through Bluetooth radio control, much like the radio control car in Chapter 7. It's a fun project that allows you to appreciate things from the perspective of a small rodent.

The robot is designed to carry the X10 camera with a battery pack, but if you use a different camera you may need to alter the plan slightly.

Design goals:

- Steerable
- Single arm in front to push LEGO ball
- Mounted video camera
- Clear view in front of robot

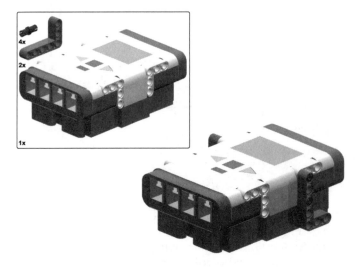

STEP 1 This robot may not look like much now, but by Step 15 it will be nearly indistinguishable from the real David Beckham.

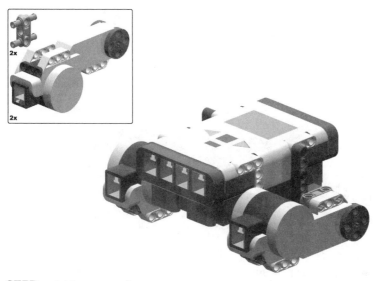

STEP 2 Add parts as shown.

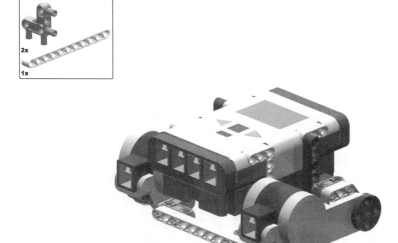

STEP 3 Add parts as shown.

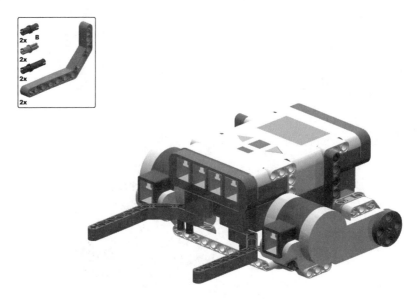

STEP 4 Insert the blue pins into the axle holes on the end of the bent beams (obscured). In the hole next to them, insert the long pins as shown. Attach them to the NXT brick and then add the short pins as shown.

STEP 5 Add the beam as shown. Add the short black beam using two blue axle pins. Insert the pin into the hole and connect this to the white beam (this is a little tricky).

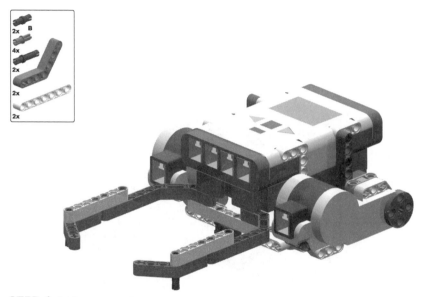

STEP 6 Add parts as shown.

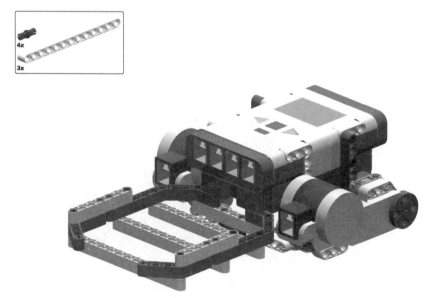

STEP 7 Add parts as shown.

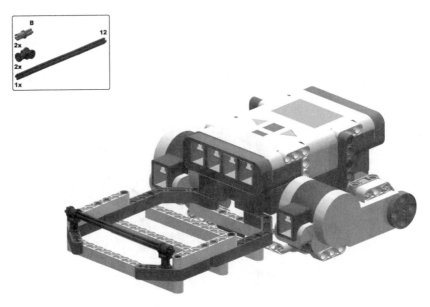

STEP 8 Add parts as shown.

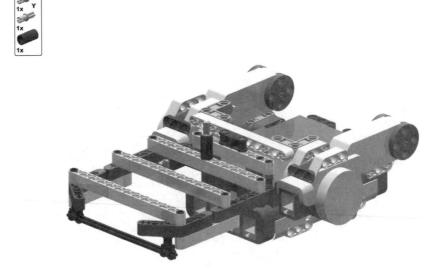

STEP 9 Flip the robot over and add the parts as shown.

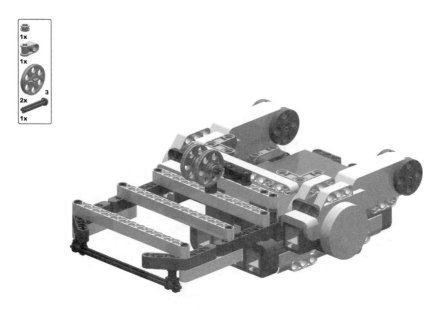

STEP 10 Place one wheel on the axle, then insert through the pin hole, then insert the other wheel and finally add the half bush.

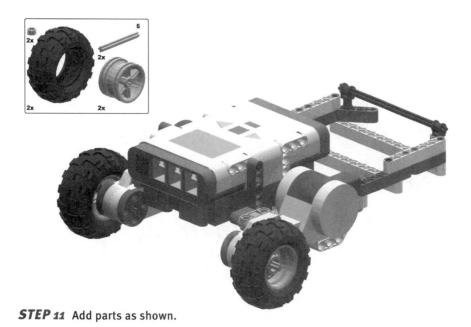

STEP 11 Add parts as shown.

STEP 12 Add parts as shown.

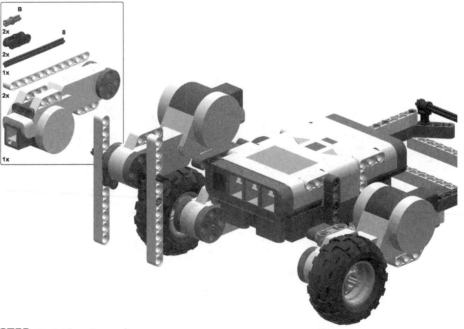

STEP 13 Add parts as shown.

STEP 14 Add parts as shown.

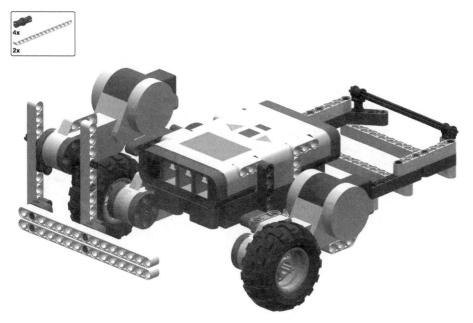

STEP 15 There. Not even David Beckham's mother could tell them apart.

While Beckhambot may not have David Beckham's kicking power, he can squeeze through tight spaces that professional footballers could not even dream of.

Connect the two drive motors to ports B and C with medium cables. Make sure to loop the cables under the robot so they don't interfere with the video image. Use a short cable to connect the kicking motor to port A. Finally, place the camera in the rear basket (see Figure 10-10).

Figure 10-10 Beckhambot ready to enter the field

Programming Beckhambot

The code for Beckambot is not very different from the code for Moon Buggy. The main difference is that Beckhambot has a kicking mechanism and a video feed. The constructor demonstrates how easy it is to initialize the video feed for display.

```
1. import icommand.vision.Vision;
2. import icommand.nxt.*;
3. import icommand.navigation.*;
4. import icommand.nxt.comm.*;
5.
6. import java.awt.event.*;
7.
8. public class Beckhambot implements KeyListener {
9.
10.    Pilot kicker = new Pilot(5.6F, 16.5F, Motor.B, Motor.C);
11.
12.    final int FORWARD = 87, // W = forward
13.    BACKWARD = 83, // S = backward
14.    LEFT = 65, // A = left turn
15.    RIGHT = 68, // D = right
16.    KICK = 75, // K = kick
17.    QUIT = 81; // Q
```

```
18.
19.     public Beckhambot() {
20.         Vision.setImageSize(320, 240);
21.         Vision.startViewer("Kick-O-Vision");
22.
23.         NXTCommand.open();
24.         NXTCommand.setVerify(true);
25.         System.out.println("BT Opened");
26.         Motor.A.setSpeed(900);
27.         kicker.setSpeed(500);
28.     }
29.
30.     public void keyPressed(KeyEvent key) {
31.         switch(key.getKeyCode()) {
32.         case FORWARD:
33.             kicker.forward();
34.             break;
35.         case BACKWARD:
36.             kicker.backward();
37.             break;
38.         case LEFT:
39.             kicker.rotate(100000, true);
40.             break;
41.         case RIGHT:
42.             kicker.rotate(-100000, true);
43.             break;
44.         case KICK:
45.             Motor.A.rotate(-180);
46.             Motor.A.rotateTo(0, true);
47.             break;
48.         case QUIT:
49.             NXTCommand.close();
50.             System.exit(0);
51.             break;
52.         }
53.     }
54.
55.     public void keyReleased(KeyEvent arg0) {
56.         kicker.stop();
57.     }
58.
59.     public void keyTyped(KeyEvent arg0) {}
60.
61.     public static void main(String [] args) {
62.         Beckhambot b = new Beckhambot();
63.         Vision.getFrame().addKeyListener(b);
64.     }
65. }
```

Results

Turn on the video receiver and run the iCommand code. You can control the robot using the keys in Table 10-2. This project is a lot of fun, especially when using only the video to control the robot. The reception is sometimes sketchy and not always clear, but good enough for our purposes.

Command	Key
Forward	W
Backward	S
Left	A
Right	D
Kick	K
Quit	Q

Table 10-2 Key commands

TRY IT: Add ultrasonic radar to your robot so you can see objects around the robot, even when it is in a different room.

Processing Visual Data

The leJOS vision API allows detection of certain properties on the screen. Currently leJOS can detect color, movement, and light. It can tell you the region where a specific property is detected. In this section we will program the Beckhambot to look for a colored ball using the video camera, move towards the ball, and attempt to kick it.

The Vision API allows you to create rectangular regions on the screen to identify properties such as color. To track the ball, we will define four regions (see Figure 10-11). The region in the center of the video tells the robot when the colored ball is directly ahead of the robot (and therefore when to move forward). As long as it detects the color in region 2, it will drive forward. If it detects the ball in a region next to center, it knows it should rotate slightly until the color is only in the center region. Region 4 is at the bottom of the screen. When the robot is close to the ball, it will only show up in region 4 (and not region 2). This means it should kick the ball.

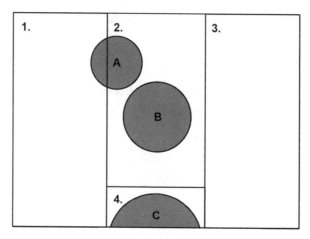

Figure 10-11 Tracking the ball in regions

Since the robot plays soccer automatically, we'll call it Autobot to differentiate it from Beckhambot. The code for Autobot is as follows:

```
1.  import icommand.navigation.*;
2.  import icommand.nxt.Motor;
3.  import icommand.nxt.comm.NXTCommand;
4.  import icommand.vision.*;
5.
6.  public class Autobot implements ColorListener {
7.
8.      Pilot kicker = new Pilot(5.6F, 16.5F, Motor.B, Motor.C);
9.      long lastOne;
10.
11.     private final int COLOR = 0xa0b0c0;
12.
13.     public static void main(String [] args) throws
            Exception {
14.         Autobot ab = new Autobot();
15.         ab.init();
16.
17.     }
18.
19.     private void init() {
20.         Vision.setImageSize(320, 240);
21.         Vision.addRectRegion(1, 0, 0, 100, 240);
22.         Vision.addColorListener(1, this, COLOR);
23.         Vision.addRectRegion(2, 100, 50, 120, 190);
24.         Vision.addColorListener(2, this, COLOR);
25.         Vision.addRectRegion(3, 220, 0, 100, 240);
26.         Vision.addColorListener(3, this, COLOR);
27.         Vision.addRectRegion(4, 100, 0, 120, 50);
28.         Vision.addColorListener(4, this, COLOR);
29.         Vision.startViewer("Kick-O-Vision");
```

```
30.          kicker.rotateLeft(); // begin search for ball
31.      }
32.
33.      public void colorDetected(int region, int color) {
34.          if ((System.currentTimeMillis() - lastOne) > 200) {
35.              lastOne = System.currentTimeMillis();
36.              switch(region) {
37.              case 1:
38.                  kicker.steer(25);
39.                  break;
40.              case 2:
41.                  kicker.forward();
42.                  break;
43.              case 3:
44.                  kicker.steer(-25);
45.                  break;
46.              case 4:
47.                  kicker.stop();
48.                  Motor.A.rotate(-180);
49.                  Motor.A.rotateTo(0, true);
50.                  break;
51.              }
52.          }
53.      }
54. }
```

Results

The kicking mechanism partially blocks the view on the right side, occasionally making Autobot blind to the ball (see Figure 10-12). Approaching the ball from the left works better; this is what the code does.

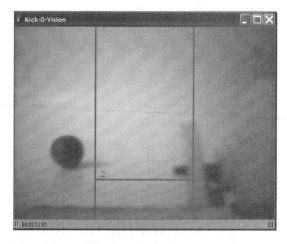

Figure 10-12 Robot-view as it lines up the ball

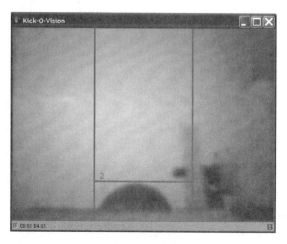

Figure 10-13 Robot view as it is about to kick the ball

In practice, this robot has a difficult time detecting the ball when it is far away. The ball is too small to have a big impact on the color average for a region. The robot really needs to be close to the ball to identify it (see Figure 10-13).

NOTE: *The Vision API detects color change by averaging the color levels and detecting any change from those averages. This could result in effects caused by the color of your walls and the environment. For example, a room with blue walls will not be able to detect a blue ball very well. For best results, you should run this project in a room with white walls and a light colored floor.*

Standing Tall

Topics in this Chapter

- Theory of Balance
- Walking
- EDDY-209
- Tilt Sensing
- Sejway

Chapter **11**

alance is something we humans take for granted because we are
so good at it. Most animals walk on four or more legs, giving
them chair-like stability when at rest. A chair maintains static
balance; it doesn't have to actively reposition itself to remain upright.
We know that a chair with only two legs tips over, yet humans are able
to stand with two legs. This is because we maintain *dynamic balance*
by constantly shifting our bodies to avoid falling. This chapter explores
static and dynamic balance concepts.

Theory of Static Balance

Static balance applies to objects at rest. Even when a robot moves,
however, static balance concepts can still be used to determine if the
robot will remain stable. For example, a vehicle with four wheels does
not tip over while it is stationary or while moving. The same principle
that gives it stability at rest also gives it stability while in motion.

Chapter six emphasized that it is desirable to build a stable robot;
robots that constantly tip over are not usually considered successes.
This section will tell you why a robot is unstable and how to make
changes to make it stable.

Every part of a rigid body is composed of particles. Gravity exerts
a downward force on each particle. The sum total of the force on each
particle produces a single direction of force, called the center of gravity,
or center of mass (see Figure 11-1). The center of gravity is another term
for center of mass when operating in a uniform gravity field. Since all
our experiments are taking place on earth, we can use the term center
of gravity.

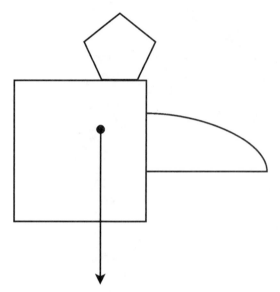

Figure 11-1 Center of Gravity (side view)

To maintain stability, you must keep the center of gravity within the center of a polygon defined by the wheels or legs of your robot. Each part of a robot that touches the ground is a point, so a robot with four wheels defines a rectangle with four points. To maintain stability, the center of gravity must reside within these points (see Figure 11-2A). If the center of gravity is outside these points, the robot will tip over (see Figure 11-2B).

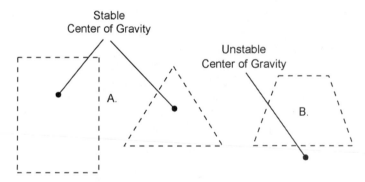

Figure 11-2 Stable and unstable center of gravity (top view)

I've sometimes built robots that were stable when stationary but unstable while moving. The three wheel robot in Figure 11-3A is an example. The center of gravity was too close to the boundary and it was a top-heavy robot. When the robot accelerated or decelerated, it introduced momentum, which shifted the center of gravity outside the polygon formed by the wheels (Figure 11-3B).

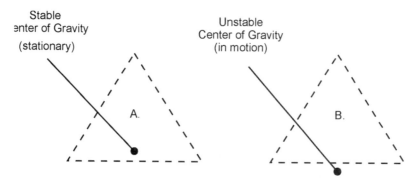

Figure 11-3 Stable while at rest, unstable in motion

Momentum is a force, just like gravity. It is calculated by multiplying mass and velocity. An object at rest has zero velocity, therefore it has no momentum.

Gravity is a force, determined by acceleration (9.8 ms² towards the ground) and mass. Momentum is a force determined by the velocity of your robot and the direction of travel. So when these forces are combined, it effectively moves the center of gravity.

The full equation to determine whether your robot will tip over is very complicated and beyond the scope of this book. It would have to take into account the height of the robot, the mass at the top of the robot versus mass at the bottom, and the sizes and angles of its various parts. Instead, we can break this down into some simple rules for stability:

- Top heavy robots are less stable than bottom heavy robots
- Taller robots are generally less stable than low robots
- Robots with high acceleration are more likely to tip over than robots with low acceleration
- Robots with mass centered over their base are more stable

Now that you know something about center of gravity, let's explore how this will affect bipedal robotics.

Bipedal Locomotion

Birds, man, and some lizards are bipedal. Flight is the primary form of locomotion for birds, and bipedal locomotion is their secondary form (barring ostriches), leaving man and some lizards as the only animals with primary bipedal locomotion. In other words, bipedal locomotion is an unpopular form of travel in nature, probably because it isn't optimal. Despite this, robot enthusiasts are drawn to creating bipedal robots; it allows us to mold creations that are similar to ourselves.

In mammals, the most popular base for a leg is the paw, which contains about five cushy pads and retractable claws, making them robust for gripping different surfaces. The least popular base is the foot, which offers bipedal creatures additional balance.

Most LEGO walking robots use two motors, because of the low number of motor ports on MINDSTORMS bricks. Each motor allows a joint to move back and forth along one plane, otherwise called the *axis of rotation.*

Now look at one of your own legs. Your leg has three active joints: the hip, knee, and ankle joints. The hip has a ball joint and can move within about a 90 degree range in any direction, forming a cone of movement (see Figure 11-4). The purpose of this movement is to allow your leg to stop you from falling over should you start to lean in the wrong direction. If you tilt in any of the 360 degrees while standing on one leg, the other leg can move to the opposite point to reestablish the center of gravity. Your hip joint allows more movement in the forward direction than the backward direction, which makes it easier for you to travel forward than backward.

Figure 11-4 Hip joint

TRY IT: *Stand up and keep your legs stiff. Concentrate on your hip joint and try to keep your leg moving only along one plane, like a robot. To help you do this, keep your arms by your side. Furthermore, lock your heel so you are not using your foot for balance. Also, and this is the hardest one to do, don't tilt your pelvis by rotating your hip socket sideways; in other words, keep your pelvis horizontal with the floor. Notice how you have to scrape your foot along the ground because you can't lift it using your hip sockets? The best you can do is shuffle by leaning back and forth with your upper torso and tilting your torso side to side. Likely your legs will have to break one plane of rotation to maintain balance. It's not so easy to walk on two joints, is it? This is the problem presented by having only three motors in the NXT kit.*

The knee is different from the hip joint because it can only rotate along a single plane. The knee allows you to walk up stairs or other platforms. When you walk, your knee bends in coordination with the hip, allowing you to fold up your leg so it doesn't scrape the ground, then reposition your foot at a different level.

TRY IT: *Try walking up some stairs using only your hip joints, as described in the last 'Try it'. Obviously you can't. Now try walking up the stairs by allowing your knee to raise your leg. Study this motion, thinking of ways you can create it in robotics.*

Your ankle joint is somewhat like your hip joint in that it can rotate along more than one plane, though it is more limited. This allows you to place your foot parallel to the surface you are traveling on. The ankle joint is more important for standing in one place than for moving, though it can position the foot to absorb shock when you are moving. (Without feet you would have a very jarring run.)

TRY IT: *Try standing on one foot with your other leg raised about six inches off the ground in front of you. While doing this, watch the ankle of the foot that is still on the ground (take off your shoe and sock so you can see your ankle at work). Notice that the ankle is making very small movements? Your ankle is constantly moving your center of gravity from one point to another to keep you from falling. If you move your arms or if there is some wind, your ankle adjusts the center of gravity to compensate.*

A shuffler is the most common walker to build with a limited number of motors. Alpha-Rex is a great example of this type of robot. It uses sophisticated engineering, allowing the ankles to move along two axes, just like your own. The downside is that it doesn't use interesting or sophisticated programming.

In recent years, companies have produced impressive bipedal robots. For example, Futaba, makers of radio control systems, introduced a bipedal robot with impressive movement capabilities (see Figure 11-5).

Figure 11-5 Futaba's robot (copyright 2007 Futaba)

This robot uses 20 servo motors to achieve lifelike movement (11 high torque servos and 9 lightweight servos). Obviously a robot like this is beyond the capability of the three servos in the NXT kit. Even with 20 servo motors, the robot still turns poorly, shimmying around to turn in a circle. It moves much the same as Alpha-Rex.

This chapter will explore walking robots using the limited number of motors in the NXT kit. These robots will attempt to lift the foot off the ground, rather than shuffle.

EDDY-209

Some robots that walk are more like windup toys. The question of how to make them walk is purely mechanical with no real programming challenge. You don't even need a processor as no real calculations take place. Other than the engineering design, there is nothing intellectually interesting about them. At the same time, there is no way to make a robot as interesting as the Futaba with just the NXT kit.

EDDY-209, the robot in this section, is somewhere in the middle. It lifts its leg off the ground and can walk with deliberate steps, making it an improvement over a shuffling robot. It keeps track of leg position and head tilt, and the programming ensures that it moves within boundaries that will not cause it to tip over.

However, it can't walk over books or climb small stairs and it does not sense very much about its current state. Specifically, it doesn't sense if a foot is on the ground and it can't tell if it has tipped over. It also doesn't perform calculations to determine a stable center of gravity.

EDDY-209 is a mean machine, inspired by ED-209 from the movie Robocop (1987, Orion Pictures). I like the way ED-209 shifts its body from right to left in order to walk, so I copied this in the design. As you will discover, neither ED-209 nor EDDY-209 are very good with stairs.

Building EDDY-209

There are three main mechanisms that allow EDDY-209 to walk. The first is the head, or NXT brick. This was used because the batteries are very heavy – enough to cause the robot to tilt. When the brick moves to the extreme left or right, the whole robot tilts to the side and one leg lifts into the air (see Figure 11-6A).

The second walking mechanism is the hip joint. Unlike your own hips, these hips rotate only along one axis. When a leg is lifted off the ground the opposite hip rotates, causing the raised leg to swing around to a new location (see Figure 11-6B).

The ankles are the third walking mechanism of the design. The ankles are not motorized so they can't be used to dynamically reposition the center of gravity like human ankles. It just acts as a fulcrum to see-saw one of the legs off the ground.

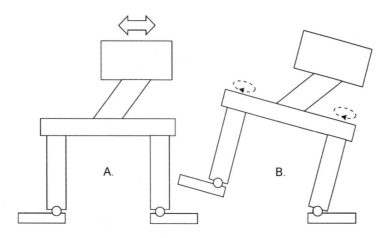

Figure 11-6 EDDY-209

The foot itself is an important aspect of the robot. It must be large enough to keep the center of gravity within the outer points of the foot (see Figure 11-7A). This is further complicated by the movement that occurs when the foot swings around to a new location. When this happens, the center of gravity moves forward, potentially toppling the robot onto its nose. In order to compensate for this, the foot is large and long.

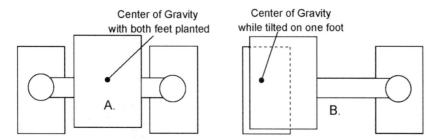

Figure 11-7 EDDY-209's Center of Gravity (top view)

When building EDDY-209, I wasn't sure where all the balance points were located, so I tried to make every aspect of the robot adjustable. The ankle-tilt can be selected by moving the crossbar back and forth. It needs to tilt enough so the opposite leg lifts off the ground, but not so much that the robot falls over. The head (NXT) needs to be able to over-shift so it can always make the robot tilt. Rotation joints can spin all the way around, even though we won't use this feature. Finally, the legs can be lengthened easily.

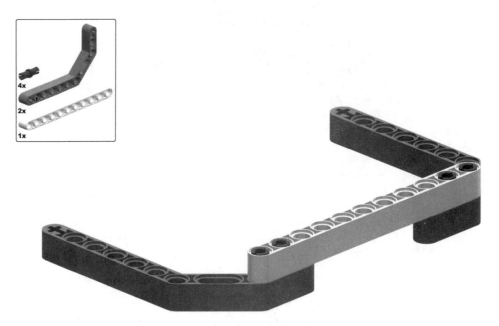

STEP 1 Add parts as shown.

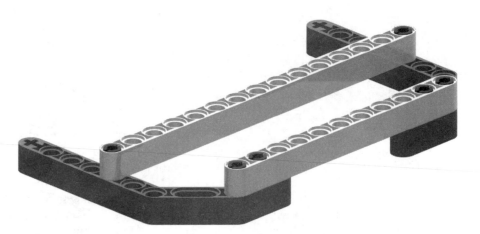

STEP 2 Add parts as shown.

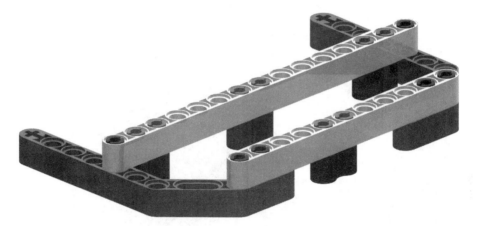

STEP 3 Add parts as shown.

STEP 4 After completing this step, repeat a mirror image for the other side.

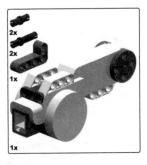

STEP 5 In this step, use a long and short pin in one side, and a short and long pin in the other.

STEP 6 Add parts as shown.

STEP 7 Add parts as shown.

STEP 8 The perpendicular axle joiner hangs loosely for now, secured by the grey axle.

STEP 9   A single black pin in the available hole holds the L-beam in place.

STEP 10   Add parts as shown.

STEP 11 Add parts as shown.

STEP 12 Add parts as shown.

STEP 13 Insert the axle into the motor. The half-bush holds the 3-unit beam in place.

STEP 14 The blue pin helps secure the top of the small L-beam.

STEP 15 Attach the closest 5-unit beam using 4 pins. The lone long-pin goes through the other 5-unit beam, the 3-unit beam, and the white beam. Use the remaining three pins to secure the 5-unit beam.

STEP 16 Add angle connectors to both sides of the leg as shown. One uses a long grey pin, since there are only two short grey pins in the kit. Now repeat a mirror image of this leg for the other side (as shown in step 38).

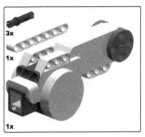

STEP 17 Add parts as shown.

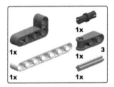

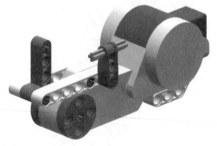

STEP 18 Add parts as shown.

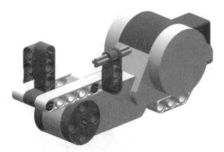

STEP 19 Attach the dark grey beam with an axle. Use a bush in the center of the motor connectors to secure it (partially obscured).

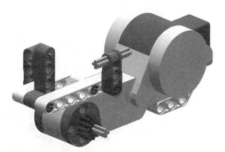

STEP 20 The bush attaches to the axle on other side of the motor (obscured).

STEP 21 The axle is held in place by a half-bush (obscured).

STEP 22 This axle holds the small L-beam in place.

STEP 23 Add parts as shown.

STEP 24 Feed the angle connector on the axle as shown. Insert the 2-unit axle in the angle connector. Attach the long-pin with bush to the 2-unit axle as shown. Add the remaining parts as shown.

STEP 25 Make sure the axles are pushed in almost to the edge of the half-bushes, otherwise this part will catch when the leg rotates.

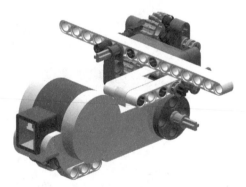

STEP 26 For now, the long beam is held in place by the black long-pin with bush and grey 3-unit axle. Use two black pins to hold the 5-unit beam to the long beam.

STEP 27 This step secures the long beam in place.

STEP 28 Add parts as shown.

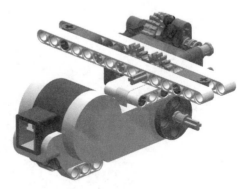

STEP 29 Attach the perpendicular joiners as shown, with the long-pins in the middle hole.

STEP 30 Attach the L-beam using a single black pin. Now insert two pins into the L-beam as shown. The perpendicular joiner attaches to the white beam using a blue axle-pin, and also the L-beam. This is tricky.

STEP 31 Secure the small gear to the axle. Attach the large gear using a yellow axle-pin.

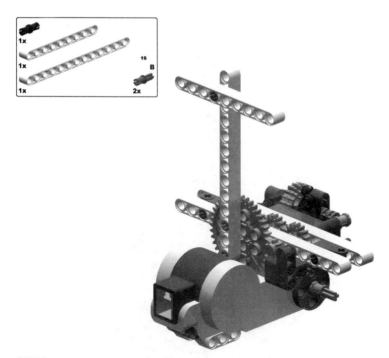

STEP 32 Add parts as shown.

STEP 33 Add parts as shown.

STEP 34 Add parts as shown.

STEP 35 Attach the white beam, as shown. Then add the perpendicular-split connectors as shown.

STEP 36 The near end of the beams attach using two long pins. The far end of the beams attach using four short pins.

STEP 37 Add parts as shown.

STEP 38 Add parts as shown.

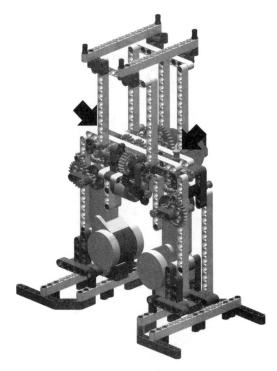

STEP 39 Attach this unit to the tops of the legs. Secure it using two bushes.

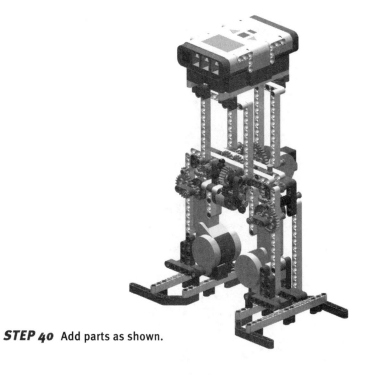

STEP 40 Add parts as shown.

Programming EDDY-209

The code for EDDY-209 performs the same series of motions repeatedly. The servo motor is the key to making this project work. The rotation sensors tell the program what position all the joints are in. Without servos, the walking algorithm would be unrepeatable. The simple code follows:

```
1.  import lejos.nxt.*;
2.
3.  public class Eddy209 {
4.
5.      public static float HEAD_TILT_ANGLE = 50;
6.      public static int STEP_DEGREES = 28;
7.      public static float GEAR_RATIO = (24F/1F); // Legs
8.      public static float HEAD_GEAR_RATIO = (40F/8F) *
            (20F/12F);
9.      public static int TILT = (int)(HEAD_TILT_ANGLE *
            HEAD_GEAR_RATIO);
10.     public static int STEP_SIZE = (int)(STEP_DEGREES *
            GEAR_RATIO);
11.     public static int PAUSE = 100;
12.
13.     public static void main(String [] options) throws
            Exception {
14.        Motor.A.setSpeed(550);
15.        Motor.B.setSpeed(800);
16.        Motor.C.setSpeed(800);
17.
18.        Motor.A.rotate((int)(TILT *1.2)); // Over tilt
             slightly
19.        Thread.sleep(PAUSE);
20.        Motor.C.rotate(-STEP_SIZE);
21.        Motor.B.rotate(-STEP_SIZE);
22.        Motor.A.rotate((int)(-TILT * 0.2));    // Correct
             overtilt
23.        for(int i=0;i<2;i++) {
24.            Thread.sleep(PAUSE);
25.            Motor.A.rotate(-TILT*2);
26.            Thread.sleep(PAUSE);
27.
28.            Motor.B.rotate(STEP_SIZE*2);
29.            Motor.C.rotate(STEP_SIZE*2);
30.            Thread.sleep(PAUSE);
31.
32.            Motor.A.rotate(TILT*2);
33.            Thread.sleep(PAUSE);
34.
35.            Motor.B.rotate(-STEP_SIZE*2);
36.            Motor.C.rotate(-STEP_SIZE*2);
37.        }
```

```
38.          Motor.C.rotateTo(0);
39.          Motor.C.flt();
40.          Motor.B.rotateTo(0);
41.          Motor.B.flt();
42.          Motor.A.rotateTo(0);
43.          Motor.A.flt();
44.      }
45. }
```

The very first half step above maneuvers the robot into the starting position. From there, it merely repeats the same movement again and again, until the loop ends. At that point, it rotates the motors back to the starting position.

Using the Robot

The robot must be positioned on a hard surface (not carpet) in the starting position, with the feet pointed forwards and the head upright. To move EDDY-209 into the starting position, you might need to hand-turn the motors. I used a connector with an axel hole in place of the half bush on the front of each motor to give my fingers something to grip.

Once it is in position, execute the program code. The robot should take a few steps and return to the starting position. Once you have confirmed that things are working well, you can try increasing the loop to make more steps, or try programming the robot to perform turns.

TRY IT: *If you own a tilt sensor, try using the sensor to determine when the robot has tilted to the side. It can react dynamically by moving the head until the tilt sensor indicates it has tilted. It can also detect if the robot has fallen over.*

Dynamic Balance Theory

Anything that falls over when left unsupported is inherently unstable. This includes objects like a broom, a two-wheeled cart, a person doing a handstand, or a motorcycle. However, we have all seen these objects balanced. How do these objects stand upright without falling over?

The theory behind dynamic balance is simple: hold a position until the object starts to fall, then shift the center of gravity to counter the direction it is tilting. Pole balancing demonstrates this concept clearly. When the pole begins to lean to the left (see Figure 11-8 A), the balancer quickly moves the base of the pole to the left, which reestablishes the center of gravity (Figure 11-8 B). When it starts to lean in another direction, the balancer shifts the center of gravity again.

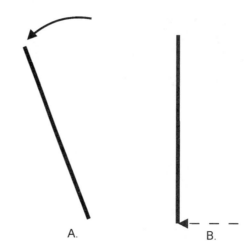

Figure 11-8 Balancing a pole

Pole balancing is complex because the pole can tip in any direction around 360 degrees. By using an axle and two wheels, we can limit the directions a robot can tip into one plane. This means the object has only the option of tipping to the left or to the right (Figure 11-9). A Segway™ illustrates this concept nicely. Let's attempt to recreate this device with LEGO.

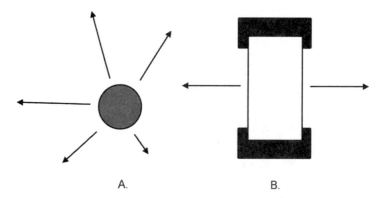

Figure 11-9 Balancing on two wheels

Building the Sejway

Sejway is an example of a balancing robot. It's a scaled down version of the Segway HT (Human Transporter). Even though the Segway uses an accelerometer to determine how to balance, the Sejway can be built with the basic LEGO NXT parts, using the light sensor to determine tilt.

The principle behind the light sensor in this robot is simple. The light sensor points at the ground with the floodlight on (the red LED light). As the robot tilts in one direction, the floodlight comes closer to the ground and the light becomes brighter, thus the light sensor detects increased values. As the robot tilts backwards the light sensor moves away from the ground and the floodlight reflection becomes weaker. In this way the LEGO NXT can detect its orientation based on the light sensor readings.

STEP 1 Add parts as shown.

STEP 2 Add parts as shown.

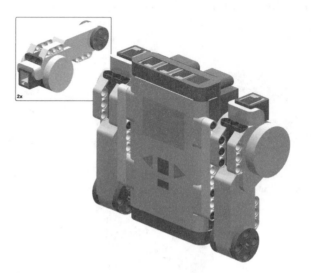

STEP 3 Add parts as shown.

STEP 4 Add parts as shown.

STEP 5 Add parts as shown.

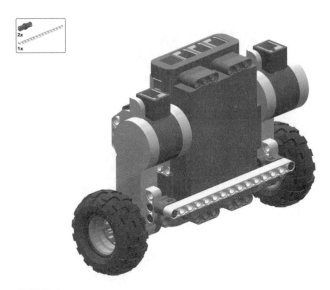

STEP 6 Add parts as shown.

STEP 7 Add parts as shown.

STEP 8 That's all!

Connect a short cable from the light sensor to port 1. Ports B and C connect to the motors (it doesn't matter which) using medium cables. Wrap the cables around the NXT to manage the excess length.

Programming Sejway

Dynamic balance is one area of robotics that needs high-speed code execution. The robot must react to changes immediately or it will lose its balance and fall over. Unlike three- and four-wheeled robots, this robot must have a fast processor and fast code. Fortunately, leJOS and the NXT are more than up to the challenge.

A *control loop* is an engineering term for a machine that reads a value, reacts according to the value, and then repeats. The sensor has a target value that it continuously tries to attain, and any difference between the current value and the target value is called *error* (see Figure 11-10).

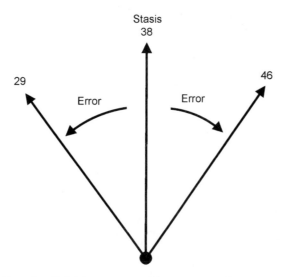

Figure 11-10 Error in a light sensor control loop

In this example, our control loop reads the light sensor value, determines which way it has tilted, and attempts to use the motors to compensate in the other direction. Control loops in engineering are famous for sometimes overreacting and causing the machine to behave erratically. This project is no different. Try the following code and observe how it overreacts:

```
1.  import lejos.nxt.*;
2.
3.  public class SejwayBad {
4.      public static void main (String[] args) {
5.          int NEUTRAL = 37;
6.          int speed = 900;
7.          LightSensor s = new LightSensor (SensorPort.S3,
               true);
8.          Motor.B.setSpeed (speed);
9.          Motor.C.setSpeed (speed);
10.
11.         while (!Button.ENTER.isPressed()) {
12.             int SENS_VAL = s.readValue();
13.
14.             boolean forward = SENS_VAL > NEUTRAL;
15.             if (forward) {
16.                 Motor.B.forward();
17.                 Motor.C.forward();
18.             } else {
19.                 Motor.B.backward();
20.                 Motor.C.backward();
21.             }
22.         }
23.
24.         s.setFloodlight(false);
25.         Motor.B.flt();
26.         Motor.C.flt();
27.     }
28. }
```

In this example, the robot goes forward when the light value is greater than 37 and backward when less than 37. This robot is very unstable and moves violently back and forth. Notice that the robot goes too far one way, and then goes too far the other way repeatedly? It doesn't learn that it is overshooting the mark each time and eventually it just falls over. We need to do something to control this behavior.

In this project we will use something called a PID Controller, often called a predictive controller. PID control uses a feedback control loop, much like the above example, but it calculates how much to react. It is also self adjusting, so if it overcorrects one time it will adjust so it doesn't correct as much on the next loop.

There are three parts to the PID control algorithm, each with a name: proportional, integral, and derivative. These three parts give the PID algorithm its name. Proportional measures the current error and determines how much it should adjust to achieve balance. Integral determines the amount of time that the error went uncorrected. Derivative is the predictive part of the algorithm that anticipates future error.

Proportional	The current error.
Integral	The amount of time that went uncorrected.
Derivative	Anticipate future error from rate of change over time.

Table 11-1 The three parts of the PID control algorithm

The PID control algorithm observes what happens after each correction. When it senses that it is no longer upright it corrects itself. Then on the next loop it checks to see if it has over corrected or under corrected. If it has overshot the target, it adjusts the derivative so it uses less power the next time. After a few tries it settles in nicely, using just enough power each time to upright itself to the balanced position. This is a form of reinforcement learning.

```
1.  import lejos.nxt.*;
2.
3.  public class Sejway {
4.
5.      // PID constants
6.      static final int KP = 28;
7.      static final int KI = 4;
8.      static final int KD = 33;
9.      static final int SCALE = 18;
10.
11.     // Global vars:
12.     int offset;
13.     int prev_error;
14.     float int_error;
15.
16.     LightSensor ls;
17.
18.     public Sejway() {
19.         ls = new LightSensor(SensorPort.S2, true);
20.         Motor.B.regulateSpeed(false);
21.         Motor.C.regulateSpeed(false);
22.     }
23.
24.     public void getBalancePos() {
25.         // Wait for user to balance and press orange
                button
26.         while (!Button.ENTER.isPressed()) {
27.             // NXTway must be balanced.
28.             offset = ls.readNormalizedValue();
29.             LCD.clear();
30.             LCD.drawInt(offset, 2, 4);
31.             LCD.refresh();
32.         }
33.     }
```

```
34.
35.    public void pidControl() {
36.        while (!Button.ESCAPE.isPressed()) {
37.            int normVal = ls.readNormalizedValue();
38.
39.            // Proportional Error:
40.            int error = normVal - offset;
41.            // Adjust far and near light readings:
42.            if (error < 0) error = (int)(error * 1.8F);
43.
44.            // Integral Error:
45.            int_error = ((int_error + error) * 2)/3;
46.
47.            // Derivative Error:
48.            int deriv_error = error - prev_error;
49.            prev_error = error;
50.
51.            int pid_val = (int)(KP * error + KI * int_error
                    + KD * deriv_error) / SCALE;
52.
53.            if (pid_val > 100)
54.                pid_val = 100;
55.            if (pid_val < -100)
56.                pid_val = -100;
57.
58.            // Power derived from PID value:
59.            int power = Math.abs(pid_val);
60.            power = 55 + (power * 45) / 100; // NORMALIZE
                    POWER
61.            Motor.B.setPower(power);
62.            Motor.C.setPower(power);
63.
64.            if (pid_val > 0) {
65.                Motor.B.forward();
66.                Motor.C.forward();
67.            } else {
68.                Motor.B.backward();
69.                Motor.C.backward();
70.            }
71.        }
72.    }
73.
74.    public void shutDown() {
75.        // Shut down light sensor, motors
76.        Motor.B.flt();
77.        Motor.C.flt();
78.        ls.setFloodlight(false);
79.    }
80.
81.    public static void main(String[] args) {
```

```
82.        Sejway sej = new Sejway();
83.        sej.getBalancePos();
84.        sej.pidControl();
85.        sej.shutDown();
86.    }
87. }
```

WEBSITE: *The above explanation is a cursory overview of a PID control loop. If you would like a full explanation of PID control, Wikipedia has an excellent article describing the equations.* http://en.wikipedia.org/wiki/PID_controller

Using Sejway

To use Sejway, run the program and place the robot on a surface that is light in color and relatively free of patterns. Set the Sejway on the wheels and position it so it is roughly balanced. Once you are sure it is in the balance position, press the orange button. The Sejway will take over and try to maintain balance.

Although two light sensors would work better than one, the robot works remarkably well. I used this robot on a white piece of paper. The longest it remained upright for over a full minute before I turned it off.

Sometimes it balances for a while then zooms away in one direction. I noticed when it was situated where a shadow falls on the paper (from a window) it would stay at that location and not zoom away. If you have problems, try to get the balancing point just right at the start.

WARNING: *You must have fully charged batteries. If your batteries are weak the NXT won't be able to move the motors fast enough to maintain balance.*

TRY IT: *Alter the code so that the Sejway acts as a regular robot—that is, it can move forward, backward, and turn left and right on command. You will have to make commands for each of these four movements:* forward(), backward(), left(), and right().

TRY IT: *If you own a tilt sensor, try it in Sejway in place of the light sensor. A tilt sensor version of the Sejway will be much more robust than the light sensor version because it is not affected by light levels or floor patterns.*

Localization

Topics in this Chapter

- Position tracking using a tachometer
- Position tracking with a compass
- Basic Navigator
- Handheld digital compass
- Compass Navigator

Chapter 12

Navigation is a key robotics concept. Without proper navigation techniques, robots would wander aimlessly. The definition of navigation is to *manage or direct the course of*. There are several aspects of managing the course of a robot:

- Localization – Where am I?
- Map Making – Where have I been?
- Path-finding – How do I get there?
- Mission Planning – Where am I going?

The next three chapters will deal with the different aspects of navigation. Before you can handle any of the last three concepts above, you need to answer the basic question: where am I? This chapter will examine ways to answer that question. We will try navigation using the parts included in the NXT kit, then attempt to increase accuracy by using a compass.

Localization Theory

Localization is the ability to determine where you are. In human language, we can describe our location in words: "I am in the living room. You are at the corner of 5th Avenue and 3rd Street. He is in Kansas." However, these types of descriptions mean nothing to a robot, since it can't understand words. Instead, your robot will use a Cartesian *coordinate system* to describe location.

A two-dimensional coordinate system keeps track of two numbers, x and y. Numbers grow larger and smaller along the x and y axes. Both of these axes start at zero and include positive and negative numbers. The x and y axes divide the system into four quadrants (Figure 12-1). Any *point* in a two dimensional area can be plotted on this grid using values of x and y.

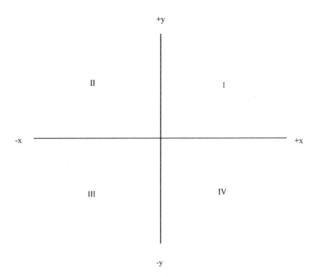

Figure 12–1 A Cartesian coordinate system

Localization can be determined in a number of ways. Some methods determine their position relative to fixed landmarks in the environment. Humans can do this by looking at local landmarks, such as your neighborhood supermarket, a tree, street signs, or any number of notable objects. GPS uses this method by noting a position relative to three or more satellites. These methods don't worry about where you have been, just where you are at a particular moment.

You can also determine location by keeping track of every movement you make starting from a point of origin. This is called dead reckoning (also sometimes called orienteering), a method of navigation used to some degree by all animals, including humans. The art of dead reckoning has been refined by sailors, geologists, forest rangers, and hikers. The only information needed for dead reckoning is direction and distance.

Direction is usually obtained using a magnetic compass. However, direction can also be determined by recording the rotation of wheels using a tachometer. For example, when one wheel rotates forward and one rotates backward, the robot rotates. Be measuring the amount each wheel rotated we can determine the angle it rotated. Distance (odometry) can also be determined by keeping an accurate record of wheel rotations.

There are some advantages and disadvantages to using this type of navigation for a robot. With round wheels and a good tachometer, a robot will outperform a human in estimating the distance it has traveled (at least on a smooth floor). The accuracy of calculating and storing coordinate points is also better in robots than in humans, although a pad of paper often helps humans to improve memory recall.

There is a downside to robots, however. Most robots are unable to self-correct their course by analyzing a situation. The robots we are making have no ability to visually recognize a target or landmark. Lastly, the NXT kit does not include a compass to help with navigation (although you can obtain one from the LEGO).

To update coordinates, the robot must perform trigonometry calculations. Luckily for us, leJOS NXJ comes with Navigator classes that do the math. However, if you would like to find out how to perform these calculations yourself you can find a simple explanation in Appendix B.

The Navigator API

The leJOS NXJ Navigator API provides a convenient set of methods to control a robot. There are methods for moving to any location and controlling the direction of movement. A Navigator object will automatically calculate angle and x, y coordinates following every move.

The great part about this class is that it can be used by any robot with *differential steering*, regardless of its construction. Differential steering has one requirement: the robot must be able to turn within its own footprint. That is, it must be able to change direction without changing its x and y coordinates. The robot wheels can be any diameter because these physical parameters are addressed by the constructor of the Navigator class. Once these parameters are set, the Navigator class works the same for all differential robots. Let's examine the actual interface.

lejos.navigation.Navigator

- `public void setSpeed()`

    Sets the speed for the robot wheels, in degrees per second.

- `public void setPosition(float x, float y, float directionAngle)`

    Allows you to reset the x-y coordinates and angle to zero, erasing the current coordinates and angle.

- `public void forward()`

    Moves the robot forward until stop() is called.

- `public void backward()`

    Moves the robot backward until stop() is called.

- `public void travel(int distance)`

    Moves the robot a specific distance. A positive value moves it forward and a negative value moves it backward. The method returns when movement is done.

    Parameters
    `distance`: The distance to travel in inches or centimeters.

- `public void travel(int distance, boolean immediateReturn)`

  Same as above except it has the option of returning immediately from the method call.

  Parameters
  `distance:` The distance to travel in inches or centimeters.

  `immediateReturn:` If true, the method returns immediately.

- `public void stop()`

  Halts the robot and calculates new x-y coordinates.

- `public void rotate(float angle)`

  Rotates the robot a specific number of degrees, in a positive or negative direction (+ or -). The angle can be any positive or negative integer, e.g. rotate (720) will cause two complete rotations counter-clockwise. This method returns only once the rotation is complete.

  Parameters
  `angle:` The angle to rotate in degrees. A positive value is counter-clockwise, and a negative value is clockwise.

- `public void rotate(float angle, boolean immediateReturn)`

  Same as above except it has the option of returning immediately from the method call.

  Parameters
  `angle:` The angle to rotate in degrees. A positive value is counter-clockwise, and a negative value is clockwise.

  `immediateReturn:` If true, the method returns immediately.

- `public void rotateTo(float angle)`

  Rotates the robot to point in a specific direction. It will take the shortest path necessary to rotate to the desired angle. The method returns once rotation is complete.

  Parameters
  `angle:` The angle to rotate to, in degrees.

- `public void rotateTo(float angle, boolean immediateReturn)`

  Same as above, except it has the option of returning immediately from the method call.

  Parameters
  `angle:` The angle to rotate to, in degrees.

  `immediateReturn:` If true, the method returns immediately.

- `public void goTo(float x, float y)`

  Moves to any point on the coordinate system. The method rotates the robot towards the target coordinates and travels the required distance. The stop() method can be called at any time to stop movement and recalculate the x-y coordinates.

  Parameters

  x: The target x coordinate.

  y: The target y coordinate.

- `public float getX()`

  Returns the current x coordinate of the robot. If the robot is moving it will calculate the present coordinate and return the value.

- `public float getY()`

  Returns the current y coordinate of the robot. If the robot is moving it will calculate the present coordinate and return the value.

- `public float getAngle()`

  Returns the current angle the robot is facing. If the robot is rotating it will not calculate the present angle, so make sure the robot is stopped when calling this method.

- `public boolean isMoving()`

  Indicates whether the robot is currently traveling under power.

- `public float distanceTo(float x, float y)`

  A helper method that returns the distance from the robot to the point with coordinates (x,y).

- `public float angleTo(float x, float y)`

  Another helper method that returns the angle to the point with coordinates (x,y).

  The Navigator interface is implemented by the classes Tacho-Navigator and CompassNavigator. NXT robots can have different wheel radii and axle lengths, so the TachoNavigator constructor requires this information. It can get this information directly, or via a Pilot object. The constructors for TachoNavigator are listed below:

lejos.navigation.TachoNavigator

- `public TachoNavigator(float wheelDiameter, float trackWidth, Motor leftMotor, Motor rightMotor)`

  Allocates a Navigator object and initializes if with the left and right wheels. The x and y values will each equal zero (centimeters) on initialization, and the starting angle is zero degrees.

  Parameters
  `wheelDiameter`: The diameter of the wheel, usually printed right on the tire, in centimeters (e.g. 49.6 mm = 4.96 cm = 1.95 in)

  `trackWidth`: The distance from the center of the left tire to the center of the right tire, in units of your choice

  `rightMotor`: The motor used to drive the right wheel. e.g. Motor.C

  `leftMotor`: The motor used to drive the left wheel. e.g. Motor.A

- `public TachoNavigator(float wheelDiameter, float trackWidth, Motor leftMotor, Motor rightMotor, boolean reverse)`

  Identical to the above constructor, except it reverses the motor movement if set to true.

- `public void turn(float radius)`

  Moves the robot in a circular path with a specified radius. The center of the turning circle is on the right side of the robot if parameter radius is negative.

  `radius`: The radius of the circular path. If positive, the left wheel is on the inside of the turn. If negative, the left wheel is on the outside.

- `public void turn(float radius, int angle)`

  Moves the robot in a circular path through a specific angle. The center of the turning circle is on the right side of the robot if radius is negative. Robot will stop when total rotation equals angle. If angle is negative, robot will travel backwards.

- `public void turn(float radius, int angle, boolean immediateReturn)`

  Same as above, except returns immediately if immediateReturn is true.

NOTE: *Sometimes the motors are reversed, causing your robot to move in directions you did not intend. This is easy to correct:*

Problem	Solution
Goes backward instead of forward	Set the Boolean parameter to true in the Navigator constructor.
Robot turns right instead of left	Swap the motors in the constructor (or swap the wire inputs)

Basic Navigation

Let's try to create a robot using NXT parts that can perform some basic navigation. This robot uses the tachometers built into the NXT motors to record all movements and thus it can give an approximation of its location.

Building a Navigating Robot

The navigation robot in this chapter is named Blighbot, after William Bligh, captain of the HMS Bounty. After Bligh lost his ship to mutiny, he made one of the most remarkable voyages in a small open boat, without navigational equipment or maps, successfully bringing every one of his loyal crew back to safety. Hopefully some of that accuracy will rub off on Blighbot.

Blighbot, uses differential steering so it will be compatible with the Navigator classes. The simple construction mounts two motors to the NXT brick. A caster wheel mounts directly to the underside of the NXT brick.

STEP 1 Add parts as shown.

STEP 2 The short end of the axle is downward.

STEP 3 Place a wheel on the axle, then insert into the caster, then add the other wheel and a half bush.

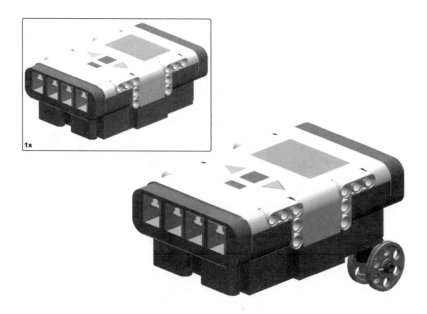

STEP 4 Attach the caster to the underside of the NXT brick.

STEP 5 The blue pins are inserted into the bent liftarm, then into the top holes of the NXT brick.

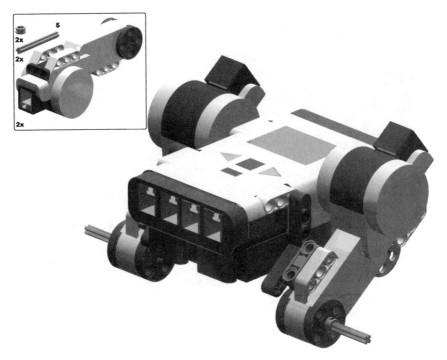

STEP 6 Add parts as shown.

STEP 7 Add parts as shown.

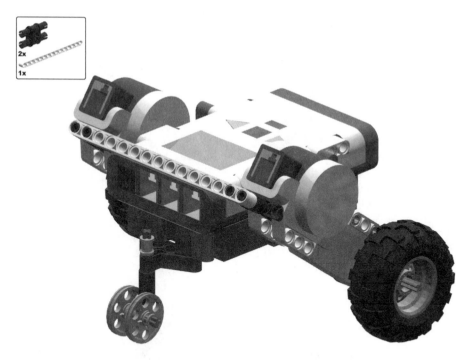

STEP 8 Turn the robot around and insert the rear support.

STEP 9 Insert medium cables into port B and C. Wrap them below the NXT and around into the motors to take up slack.

Programming Blighbot

The TachoNavigator class requires three basic parameters to work properly:

- Tire Diameter
- Track Width
- Motors

Tire diameter is the widest measurement from one side of a tire to the other. This is easy to acquire because LEGO prints the diameter right on the rubber. The wheels in the NXT kit are all 56 mm (5.6 cm).

Track width is the measurement from wheel to wheel. There should be a theoretical point where the wheel touches the ground, but LEGO tires are wide and squishy, which makes this difficult to determine precisely. Since LEGO tires are symmetrical, the best way is to measure from the center of one tire to the center of the other (see Figure 12-2). Blighbot has a track width of 16 cm according to my measurement.

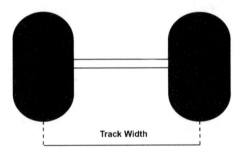

Figure 12-2 Track width

The following program puts Blighbot through an obstacle course. The challenge is to make Blighbot navigate around various obstacles and return home to the starting position (see Figure 12-3). Rather than making Blighbot trace a figure eight, he will trace two (sideways) figure fours (one without a compass and one with the compass). This course was created by laying down pennies. The reason for pennies is that if Blighbot runs over one, it is not liable to throw him off and he can continue with the rest of the course. Using this course, we can determine how accurate the Navigator classes are using only the tachometer.

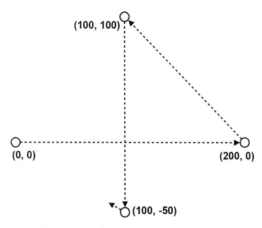

Figure 12-3 Navigating the penny obstacle course

```
1. import lejos.nxt.*;
2. import lejos.navigation.*;
3. public class Blighbot {
4.    public static void main(String [] args) {
5.       TachoNavigator robot = new TachoNavigator(5.6F, 16.0F,
             Motor.C, Motor.B, true);
6.       robot.goTo(200,0);
7.       robot.goTo(100,100);
8.       robot.goTo(100,-50);
9.       robot.goTo(0,0);
10.   }
11. }
```

Testing and Results

You need an area of about four meters square for this course. Try measuring and placing pennies at the locations shown in Figure 12-3. It's important to have Blighbot lined up with the furthest penny when you start the program. Your measurements for the other pennies are not as vital, since you can tell how accurate Blighbot is by how closely he arrives back at the start penny.

Blighbot does a good job of measuring distances but his weakness arises when he rotates. You'll notice that at waypoint 1 he is off a bit, which causes him to miss waypoint 2, and the final turn adds even more error so he really misses waypoint 3 and the trip back to the point of origin.

This problem of small errors that add up into large errors is called *drift*. The errors are cumulative, so the longer your robot travels, the further off course it becomes until the location coordinates are essentially meaningless. This is a chronic problem with tachometer rotations. You can minimize drift problems by running the robot on a smooth, hardwood floor, but you can also seek to improve rotation accuracy. The next section will attempt to make turns more accurate.

Compass

Birds are famous for the accurate navigational skills that allow them to migrate thousands of miles and arrive at the same destination every year. Birds have a magnetite crystal behind their beaks. This allows them to recognize the direction in which they are traveling. Surprisingly, man also has a magnetite crystal located behind the bridge of the nose. Unfortunately, our nose compass appears to be a vestigial remnant of our evolutionary past.

The compass can be a valuable tool in robotics. LEGO did not offer a compass sensor with the RIS kit, forcing users to build their own. This time, users can purchase a compass sensor online directly from LEGO, or from another third-party vendor (see Appendix A).

The compass has the advantage of indicating North, or in the polar coordinate system, something we can designate as zero angle. Without the compass, you just picked an arbitrary direction to indicate 0 degrees. With a compass, we can use the East heading to indicate 0 degrees on the polar coordinate system.

Calibrating your Compass

A compass is sensitive to magnetic fields. It senses the earth's magnetic field to determine North. Unfortunately, other objects in the environment, such as metal support beams, duct work, metal vent covers, nails, metal chairs, lamp stands, wiring, and even metallic deposits in the earth, produce their own magnetic fields. Even NXT motors produce magnetic fields, which can potentially distort compass readings.

The digital compasses currently on the market can be recalibrated to compensate for local magnetic fields. You don't have to recalibrate every time you use your compass, just when you move to a different location or connect the compass to a different robot where the compass might be closer or farther from motor coils. Calibration occurs by sending a start command to the compass, rotating the compass slowly two or more times, and then sending a stop command.

Attach a compass to Blighbot as shown in Figure 12-4.

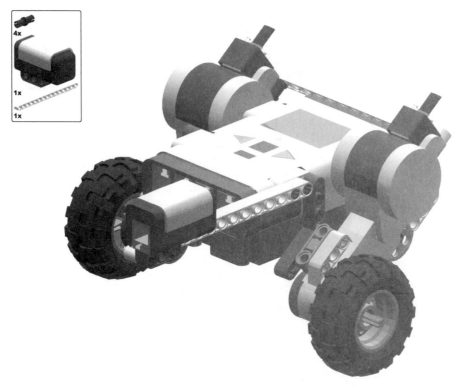

Figure 12-4 Attaching a compass to Blighbot

Now upload and run the following code:

```
1.  import lejos.nxt.*;
2.  import lejos.navigation.*;
3.
4.  public class CompassCalibrate  {
5.
6.    public static void main(String [] args) {
7.      CompassSensor cmps = new CompassSensor( SensorPort.S1);
8.
9.      Pilot hector = new Pilot(5.6F, 16.0F,Motor.B, Motor.C,
            true);
10.     cmps.startCalibration();
11.
12.     hector.setSpeed(50);
13.     hector.rotate(1000);
14.
15.     cmps.stopCalibration();
16.     hector.stop();
17.   }
18. }
```

Once this code is executed, the calibration settings are stored in non-volatile memory on the sensor, so you won't have to rerun the calibration when you turn the NXT brick off.

Handheld Digital Compass

Before we try using the compass for navigation, let's have some fun with a handheld compass. We'll use the Graphics class to draw a compass indicator to the LCD screen, and then you can walk around your house testing the effectiveness of the compass.

```
1.  import lejos.nxt.*;
2.  import lejos.navigation.*;
3.  import javax.microedition.lcdui.Graphics;
4.
5.  public class HandHeldCompass {
6.
7.    public static void main(String [] args) throws
         Exception {
8.      CompassSensor cmps = new CompassSensor(SensorPort.S1);
9.      Graphics g = new Graphics();
10.     while(!Button.ENTER.isPressed()) {
11.       g.clear();
12.       int angle = (int)cmps.getDegrees();
13.       LCD.drawInt(angle, 0, 0);
14.       g.fillArc(10,0,62, 62, angle-4, 8);
15.       g.refresh();
16.       LCD.refresh();
17.       Thread.sleep(200);
18.     }
19.   }
20. }
```

As you test the compass, you will notice that 0 is due north and angle increases clockwise. Like any compass, when you take it in the vicinity of a metallic object, the readings are affected. The compass manufacturers recommend mounting the compass at least 10 to 15 cm from the NXT and motors. In my tests, moving the compass from 7 cm to 13 cm from the NXT motors resulted in a tremendous increase in accuracy.

NOTE: *The HiTechnic/LEGO compass reads zero when the front part of the compass (the black face) points North. The Mindsensors.com compass reads zero when the small white arrow points North.*

Compass Navigation

Now it is time to test Blighbot again, only this time we will use a compass to run the obstacle course. The code is virtually identical to the previous code, except that a CompassSensor object is created and fed to the CompassPilot constructor:

```
1.  import lejos.nxt.*;
2.  import lejos.navigation.*;
3.
4.  public class BlighbotCompass {
5.
6.     public static void main(String [] args) {
7.        CompassSensor cps = new CompassSensor(SensorPort.S1);
8.        CompassPilot pilot = new CompassPilot(cps, 5.6F,
              16.0F,Motor.B, Motor.C, true);
9.        Navigator robot = new CompassNavigator(pilot);
10.
11.       robot.goTo(200,0);
12.       robot.goTo(100,100);
13.       robot.goTo(100,-50);
14.       robot.goTo(0,0);
15.    }
16. }
```

This time Blighbot performs much better. He comes to within 10 centimeters of the goal, while the previous test only came within 30 centimeters. Since the compass is relying on a fixed reference point for direction (the earth's magnetic field) it is not likely to drift as much as navigation unaided by a compass.

Although the compass reduces errors in rotation, the robot still has a drift problem. Occasionally your robot will hit an object and bounce to a totally different position. That is rare and very noticeable. However, when in motion, your robot continually hits little bumps, in the carpet or on your floor, or even tire treads. These little bumps add up into one big bump. Even with the compass, when your robot tries to rotate 15 degrees it might actually rotate 16 degrees. These small errors add up too. These problems will always plague dead reckoning, which makes landmark navigation more appealing.

In the next chapter we will examine a robot that makes a map from its environment.

Mapping

Topics in this Chapter

- Automatic Mapping
- Magellan—a robot that makes maps

Chapter 13

A map is a planar representation of a region. A robot that is capable of knowing its location, such as those explored in the previous chapter, can create a map of every place it has been. Depending on the type of sensors on the robot, it can construct different kinds of maps. If the robot is equipped with a light sensor it would be able to create a light intensity map of your home. If equipped with a temperature sensor, it would be able to create a thermal map. In this chapter, we will map solid objects using the ultrasonic sensor.

Automated Mapping

In the film *Forbidden Planet* (1956) a robot called Robby transforms into a vehicle and drives across an alien landscape. Robby has two rotating sensors on his head that he presumably uses to detect objects around him. This project uses the same concept, a rotating sensor to map the environment.

In my previous book, Core LEGO MINDSTORMS, I wanted to use a compass sensor to guide a robot that would map its surroundings. Unfortunately, after programming the sensor and the navigation code there was no memory left for mapping functions. Memory is not a problem this time; there is enough on your computer to store much more data.

Mapping Basics

When keeping track of points we use a coordinate system that can even use fractions to define a point (such as $x = 2.378$). However, to plot our map data we will use squares on a *Cartesian grid*. It's much the same idea as a coordinate system, but the basic unit in this system is *a unit square*. The square can be any size, 10 centimeters square, for instance.

In a Cartesian grid, a pair of coordinates identifies a unit square as shown in Figure 13-1. However, each pair refers to a square one unit to the right of the y coordinate, or one unit above the x coordinate. So the square indicated by 0,0 indicates a square above and to the right of the point at 0,0 (see Figure 13-1).

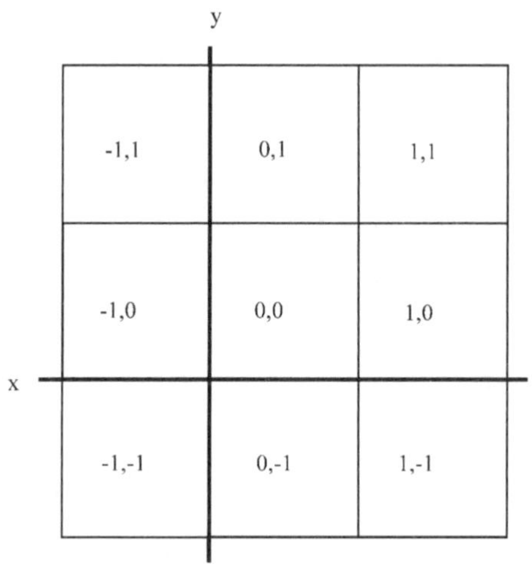

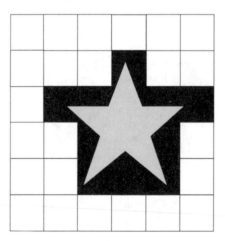

Figure 13-1 A Cartesian grid

The approach we will use for storing data is called *occupancy grid mapping.* The theory is simple. If any part of a solid object is detected within a unit square, the entire square is marked as occupied (see Figure 13-2). Thus, even if a small portion of the unit square is occupied, the robot knows it might run into an object if it attempts to enter that square.

Figure 13-2 Occupancy Grid Map

There are three possible states for each unit square on the grid:

- unknown
- occupied
- unoccupied

When the robot begins mapping a room, there is no map data so each square is marked unknown.

To store the data for each unit square, we could use 0 for unknown, +1 for occupied, and -1 for unoccupied. At times the robot might revisit squares it has previously mapped. If it scans a square twice and finds it unoccupied both times, it should indicate that it is even more confident it is unoccupied. For this we will use something called *histogramic in-motion mapping.* Although it has a big name, it simply refers to adding 1 each time it encounters an object in a square, and subtracting 1 every time it doesn't detect an object. So if it finds an object three times in a square, the square will end up with a value of +3, meaning it is very confident an object exists there. This is a type of reinforcement learning because every sweep reinforces the knowledge previously gained.

Mapping in Action

Our robot will use the ultrasonic sensor to scan for objects in the environment. The robot starts off at the arbitrary point of 0, 0 and performs a full 360 scan with the ultrasonic sensor at 30 degree intervals (12 scans in a circle), resulting in 100% coverage. All unit squares in the map data start with a value of 0. Any of the 12 scans that get a deflection will be incremented by one, while any unit squares between the outer edge of the scan and the ultrasonic sensor are decremented by one (open space).

As mentioned in chapter one, the sensor has a 30 degree scan cone. Figure 13-3 shows an ultrasonic sensor detecting an object 45 centimeters away. Because of the 30 degree scan cone, the object could reside anywhere between A and B. When mapping, we will fill in every point between A and B to be safe. This means it will produce an arc on our map for every object detected. As the sensor observes an object from different angles, the false readings should be erased, eventually producing respectable resolution for the location of objects.

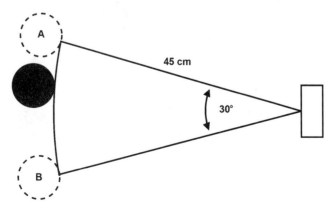

Figure 13-3 Scan cone produces multiple possibilities

The occupancy grid is made up of confidence values. The ultrasonic sensor might give wrong readings from time to time, but chances are the sonar will look at the same grid square from a number of different angles. If one time it detects an object, but the other two times it doesn't, it will end with a value of -1:

$$0 - 1 + 1 - 1 = -1$$

This example does not produce a high confidence level, but at least it correctly assumes the unit square is empty.

As we discovered in the previous chapter, the navigation coordinates drift the longer the robot travels. We must account for drift by resetting the robot periodically to ensure it has relatively accurate localization coordinates.

Exploration Strategy

There are many possible exploration strategies in mapping an area. Because this project is already very complex, I will keep the exploration strategy simple. The robot chooses the greatest distance returned to explore first. If there are many the same, it will randomly choose one direction.

Figure 13-4 shows an environment and some scans (black dots indicating coordinates). When the robot is in the left section of the room it performs a 360 degree scan at 45 degree intervals (we will use 30 degree intervals in practice). Many of the scans terminate in solid objects, while the others are nodes for potential exploration.

 NOTE: *Even though we are mapping unit-squares to a Cartesian grid, we can still use normal coordinate points to guide our robot. In effect, the coordinate points are overlaid onto the Cartesian grid.*

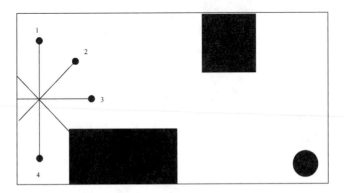

Figure 13-4 Scanning step one

The robot then chooses one of the coordinates to explore—in this case, coordinate 1. After moving to coordinate 1, it performs another scan. Most of the scans terminate at a wall (Figure 13-7), but the robot has a few more options. It then repeats.

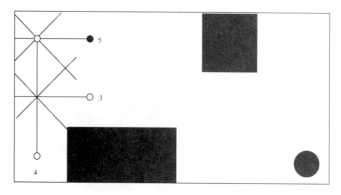

Figure 13-7 Scanning step two

In practice, this strategy might not give 100% coverage. First, because of the randomness of the exploration, it is not guaranteed to cover the entire area. Second, the ultrasonic sensor's 30 degree scan cone hinders its ability to identify small passages. But it will provide a good rough map of the area.

Building the Mapping Robot

The robot in this chapter is named Magellan, after the famous explorer. This robot uses four-wheel drive and includes a rotating ultrasonic sensor with the ability to scan 360 degrees around the robot.

STEP 1 Attach four bent-beams to the NXT.

STEP 2 Put all five black long-pins into the beam, then attach to the main unit. Place the bent beam in place and secure with the axle.

STEP 3 Repeat for the other side.

STEP 4 Reinforce the attachment by placing beams on the inside of the main unit. Attach beams on the end of the unit.

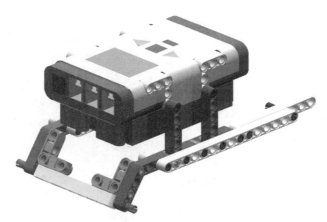

STEP 5 Add two connectors, then add the beam. Add two blue axle-pins as shown.

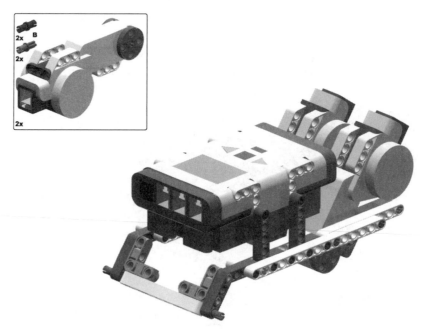

STEP 6 Add the rear supports for the motors.

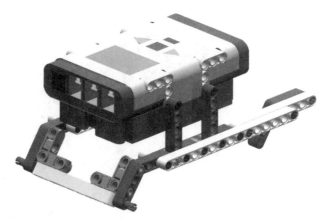

STEP 7 Attach the motors to the rear supports.

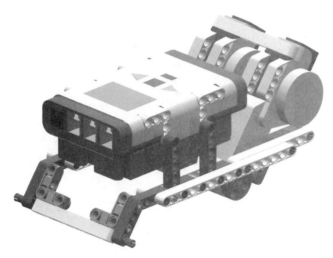

STEP 8 Secure the motors with a 9-unit beam.

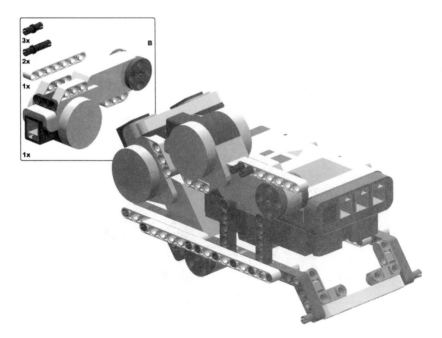

STEP 9 Attach the motor to the side of the NXT brick using three black pins. Then attach a beam to the top of the motor with long-pins.

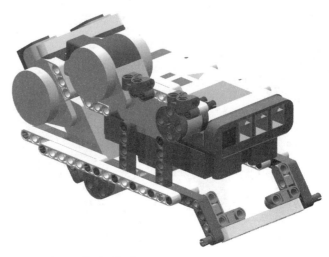

Step 10 Add two connector-units to the beam.

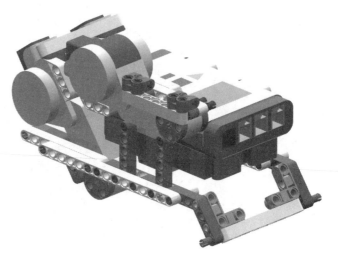

STEP 11 Secure another 7-unit beam to the underside of the connectors.

STEP 12 Attach a knob-wheel to the motor using a 4-axle.

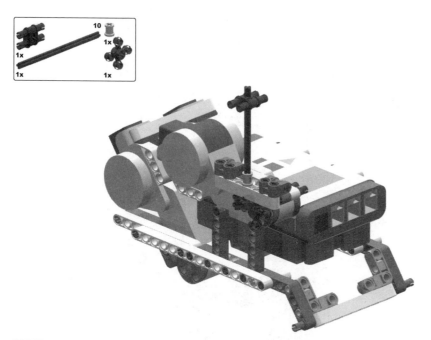

STEP 13 Mesh the previous knob-wheel with another knob-wheel and insert a 10-unit axle into it. Secure it with a bush and add the black double-pin connector.

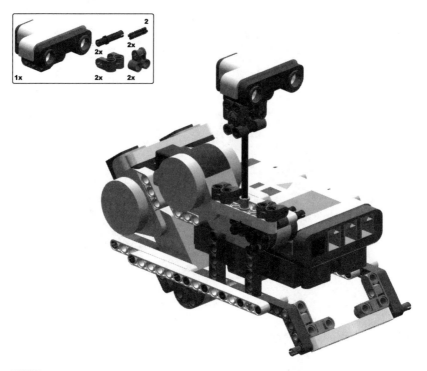

STEP 14 Add the ultrasonic sensor using connectors in the flat-orientation.

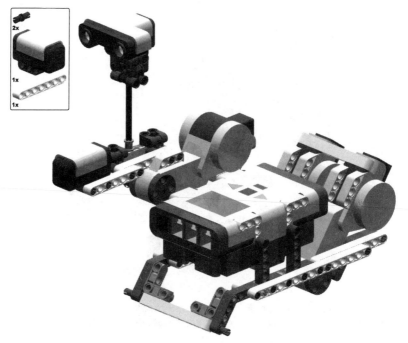

STEP 15 OPTIONAL. Add the compass sensor.

STEP 16 Start the drive unit as shown.

STEP 17 Add three idler gears. Since there are only four yellow axle-pins, use a three-unit axle and a bush in the middle.

STEP 18 Add the final axle. Position the axle in the middle.

STEP 19 Add the wheels as shown.

STEP 20 Repeat a mirror image for the other side.

STEP 21 Attach both sides to the main unit.

Using a short cable, connect port A to the motor that rotates the scanner. Connect port B to the left motor and port C to the right motor using medium cables. The cables are fed under the robot and up above the motors. Connect the ultrasonic sensor to sensor port 3 using a long cable. Connect the optional compass to sensor port 1 using another long cable.

Programming the Mapping Robot

This robot is a lot more fun if it shows the mapping data real-time on your computer screen. Rather than using leJOS NXJ, we will use the iCommand package over Bluetooth. There are more classes in this project than is usual for test code, but I've tried to keep it as simple as possible. The classes are as follows:

- MapData – Stores coordinates and values
- MapDisplay – Displays map data
- MapScanner – Calculates scan squares from ultrasonic sensor data
- Magellan – Main code for exploration

The MapData class simply stores points (x and y coordinates) along with the appropriate value for that point (0 for unknown, positive integers for solid, negative integers for open space). It also keeps track of the minimum and maximum coordinates it holds in the set.

```
1.  import java.util.*;
2.  import java.awt.Point;
3.
4.  public class MapData {
5.
6.      private HashMap <Point, Byte> points;
7.
8.      // Outer limits of map
9.      public int xMin = 0;
10.     public int xMax = 0;
11.     public int yMin = 0;
12.     public int yMax = 0;
13.
14.     public MapData() {
15.         points = new HashMap <Point, Byte> (50);
16.     }
17.
18.     public void changeValue(int x, int y, byte val) {
19.         byte oldValue = getValue(x, y);
20.
21.         points.put(new Point(x, y), new Byte((byte)(oldValue
            + val)));
22.         // Now update outer limits of map:
23.         if(x > xMax) xMax = x;
24.         if(x < xMin) xMin = x;
```

```
25.          if(y > yMax) yMax = y;
26.          if(y < yMin) yMin = y;
27.      }
28.
29.      public byte getValue(int x, int y) {
30.          Byte val = (Byte)points.get(new Point(x, y));
31.          if(val == null) val = new Byte((byte)0);
32.          return val.byteValue();
33.      }
34.
35.      public void mergeMapData(MapData other) {
36.          Point [] coords = other.getAllCoordinates();
37.          for(int i=0;i<coords.length;i++) {
38.              byte val = other.getValue(coords[i].x, coords[i].y);
39.              this.changeValue(coords[i].x, coords[i].y, val);
40.          }
41.      }
42.
43.      public Point[] getAllCoordinates() {
44.          Point [] keys = (Point [])points.keySet().toArray
                 (new Point[points.keySet().size()]);
45.          return keys;
46.      }
47. }
```

The MapDisplay accepts a MapData object and paints the coordinate data to a Canvas. The MapDisplay automatically expands the display, so if the map data is only 4 x 4, it will fill the screen with only those 16 points. This way it automatically expands to accommodate the floor space of your particular room.

```
1. import java.awt.Point;
2. import java.awt.*;
3. import javax.swing.*;
4.
5. public class MapDisplay extends JFrame {
6.
7.      private int CMS_PER_UNIT; // Used to determine size
           to draw NXT and length of ping lines
8.
9.      private MapData mapData;
10.     private MapCanvas mc;
11.
12.     public MapDisplay(MapData md, int cmPerUnit) {
13.         super("Occupancy Grid Map");
14.         this.setMapData(md);
15.         this.setSize(640, 480);
16.         mc = new MapCanvas();
17.         this.add(mc); // Might just put new MapCanvas() here!!
18.         this.setVisible(true);
19.         this.CMS_PER_UNIT = cmPerUnit;
```

```
20.      }
21.
22.      public void setMapData(MapData map) {
23.          this.mapData = map;
24.      }
25.
26.      public void update() {
27.          mc.repaint();
28.      }
29.
30.      private class MapCanvas extends Canvas {
31.          public void paint(Graphics g) {
32.
33.              // Draw box around map bL = bottom left, tR =
                    top right
34.              Point bL = toCanvasPoint(mapData.xMin,
                    mapData.yMin);
35.              Point tR = toCanvasPoint(mapData.xMax + 1,
                    mapData.yMax + 1);
36.              setBackground(Color.GRAY);
37.              g.setColor(Color.LIGHT_GRAY);
38.              g.fillRect(bL.x, tR.y, tR.x - bL.x, bL.y - tR.y);
39.
40.              // Draw squares according to mapData
41.              Point [] points = mapData.getAllCoordinates();
42.              for(int i=0;i<points.length;i++) {
43.                  Point p1 = toCanvasPoint(points[i].x,
                        points[i].y);
44.                  Point p2 = toCanvasPoint(points[i].x + 1,
                        points[i].y + 1);
45.                  int squareSize = p2.x - p1.x;
46.                  int val = mapData.getValue(points[i].x,
                        points[i].y);
47.                  if(val > 0)
48.                      g.setColor(Color.BLACK);
49.                  else if(val < 0)
50.                      g.setColor(Color.WHITE);
51.                  else
52.                      g.setColor(Color.GRAY);
53.                  g.fillRect(p1.x, p2.y, squareSize, squareSize);
54.              }
55.
56.              // Draw x axis and y axis
57.              Point origin = toCanvasPoint(0,0);
58.              g.setColor(Color.ORANGE);
59.              g.drawLine(bL.x, origin.y, tR.x, origin.y); // x
                    axis
60.              g.drawLine(origin.x, tR.y, origin.x, bL.y); // y
                    axis
61.          }
```

```
62.
63.        private Point toCanvasPoint(int x, int y) {
64.            // Actual Map size (in squares)
65.            double mapWidth = mapData.xMax - mapData.xMin;
66.            double mapHeight = mapData.yMax - mapData.yMin;
67.
68.            // Determine square size to use based on
                  mapData.xMin, xMax, yMin, xMax
69.            double canvasRatio = (double)this.getWidth()/
                  (double)this.getHeight();
70.            double mapRatio = mapWidth/mapHeight;
71.            int squareSize;
72.            // Depending on ratios, use either width or
                  height of map to determine size
73.            if(canvasRatio > mapRatio) {
74.                squareSize = (int)(this.getHeight()/mapHeight);
75.            } else {
76.                squareSize = (int)(this.getWidth()/mapWidth);
77.            }
78.
79.            int xOrigin = (int)-mapData.xMin * squareSize;
80.            int yOrigin = (int)this.getHeight() + (mapData.
                  yMin * squareSize);
81.
82.            // Calculate excess space around border:
83.            int xExcess = this.getWidth() - ((int)mapWidth *
                  squareSize);
84.            int yExcess = this.getHeight() - ((int)mapHeight
                  * squareSize);
85.
86.            int xCanvas = xOrigin + (x * squareSize) +
                  xExcess/2;
87.            int yCanvas = yOrigin - (y * squareSize)
                  - yExcess/2;
88.
89.            return new Point(xCanvas, yCanvas);
90.        }
91.    }
92. }
```

Now we need a class that can accept a distance measurement and plot these to the map display as a 30 degree cone shape. This class is called MapScanner. It doesn't use fancy algorithms like Bresenham's algorithm to plot these shapes, so doesn't produce the most refined results, but it works.

```
1. import java.awt.geom.*;
2.
3. public class MapScanner {
4.
```

```
5.    static private double SCAN_CONE = 30;  // 30 degree
          scan angle with Ultrasonic
6.    static private int MAX_SCAN = 180;  // Anything higher
          than this = no ping
7.    static private int SQUARE_SIZE = 10;
8.
9.    static public void plotSquares(Point2D.Double pos,
          double direction, int distance, MapData mapData) {
10.      // 1: Get 3 points for triangle
11.      Point2D.Double point1 = pos;
12.      Point2D.Double point2 = getPoint(distance,
            direction + SCAN_CONE/2);
13.      Point2D.Double point3 = getPoint(distance,
            direction - SCAN_CONE/2);
14.
15.      // PLOT END LINE ONLY:
16.      MapData endLineSquares = new MapData();
17.      if(distance < MAX_SCAN) {
18.          endLineSquares = MapScanner.plotLine(new
                Line2D.Double(point2, point3));
19.          // Now add the endLineSquares to md
20.          mapData.mergeMapData(endLineSquares);
21.      }
22.
23.      // Find lowest x value:
24.      double xMin = point1.x;
25.      if(point2.x < xMin) xMin = point2.x;
26.      if(point3.x < xMin) xMin = point3.x;
27.
28.      int startX = -SQUARE_SIZE/2 + (int)((xMin - pos.x)
            / SQUARE_SIZE) * SQUARE_SIZE;
29.
30.      // Find largest x value:
31.      double xMax = point1.x;
32.      if(point2.x > xMax) xMax = point2.x;
33.      if(point3.x > xMax) xMax = point3.x;
34.
35.      // Outer loop: Move one square at a time along
            graph
36.      for(int i=startX;i<xMax;i = i + SQUARE_SIZE) {
37.          // Get vertical line for this section that
                meets two parts of triangle:
38.          Line2D.Double vertical = new Line2D.Double(i,
                pos.y + distance + 1, i, pos.y - distance - 1);
39.
40.          // X position for corners of grid squares
41.          int xLeft = (int)(vertical.x1 - SQUARE_SIZE / 2);
42.          int xRight = (int)(vertical.x1 + SQUARE_SIZE / 2);
```

```
43.
44.            boolean isIntersect = trimVertical(point1,
                  point2, point3, vertical);
45.        // Now segment line into many smaller lines, if
               > 50% then add to mapData
46.        if(isIntersect) {
47.            // Inner loop: Check each square
48.            // 1 Measure from highest Y point to next
                  gridline Y ending in multiple of
                     SQUARESIZE (10)
49.
50.            int yStart = ((int)(vertical.y1 / SQUARE_
                  SIZE) * SQUARE_SIZE);
51.
52.            int yEnd = ((int)(vertical.y2 / SQUARE_SIZE)
                  * SQUARE_SIZE) + SQUARE_SIZE;
53.            if(yStart < yEnd) {
54.                int yTemp = yStart;
55.                yStart = yEnd;
56.                yEnd = yTemp;
57.            }
58.
59.            // LOOP THROUGH ALL MIDDLE SQUARES
60.            for(int y=yStart;y>=yEnd;y = y - SQUARE_SIZE) {
61.                if(endLineSquares.getValue(xLeft/SQUARE_
                      SIZE, (y - SQUARE_SIZE)/SQUARE_SIZE)
                         == 0)
62.                    mapData.changeValue(xLeft/SQUARE_SIZE,
                          (y - SQUARE_SIZE)/SQUARE_SIZE,
                             (byte)-1);
63.            }
64.
65.            // CHECK TOP SQUARE
66.            // Two corners of Y for grid squares:
67.            int yBottom = yStart;
68.            int yTop = yStart + SQUARE_SIZE;
69.            byte chVal = -1; // Value to increment
                  mapData square
70.
71.            // Now add the appropriate TOP value to
                  mapData
72.            double topPortion = vertical.y1 %
                  SQUARE_SIZE;
73.            if(topPortion / SQUARE_SIZE >= .5) {
74.                // Add this square to mapData
75.                if(endLineSquares.getValue(xLeft/SQUARE_
                      SIZE, yBottom/SQUARE_SIZE) == 0)
76.                    mapData.changeValue(xLeft/SQUARE_SIZE,
                          yBottom/SQUARE_SIZE, chVal);
77.            }
```

```
78.
79.              // Y corners of grid square:
80.              yBottom = yEnd;
81.              yTop = yEnd + SQUARE_SIZE;
82.              chVal = -1; // Value to increment mapData
                    square
83.
84.              // Add value to BOTTOM SQUARES
85.              double bottomPortion = SQUARE_SIZE -
                    vertical.y2 % SQUARE_SIZE;
86.              if(bottomPortion / SQUARE_SIZE >= .5) {
87.                 // Add this square to mapData
88.                 if(endLineSquares.getValue(xLeft/SQUARE_
                       SIZE, yBottom/SQUARE_SIZE) == 0)
89.                     mapData.changeValue(xLeft/SQUARE_SIZE,
                          yBottom/SQUARE_SIZE, chVal);
90.              }
91.           }
92.        }
93.     }
94.
95.     static MapData plotLine(Line2D.Double line) {
96.        MapData endSquares = new MapData();
97.
98.        // Get min, max X and Y points
99.        double minX = (line.x1<line.x2)?line.x1:line.x2;
100.       double maxX = (line.x1>line.x2)?line.x1:line.x2;
101.       double minY = (line.y1<line.y2)?line.y1:line.y2;
102.       double maxY = (line.y1>line.y2)?line.y1:line.y2;
103.
104.       // Start at lowest x point, move to highest x
              point.
105.       for(int x = (int)minX;x<maxX;x = x + SQUARE_SIZE-1) {
106.          Point2D.Double intersect = new Point2D.Double();
107.          Line2D.Double vertLine = new Line2D.Double(x,
                 minY, x, maxY);
108.          boolean isIntersect = getIntersection(vertLine,
                 line, intersect);
109.          if(isIntersect) {
110.             int gridX = ((int)intersect.x/SQUARE_SIZE);
111.             int gridY = ((int)intersect.y/SQUARE_SIZE);
112.             if(endSquares.getValue(gridX, gridY) == 0)
113.                 endSquares.changeValue(gridX, gridY,
                       (byte)+1);
114.          }
115.       }
116.
117.       // Now start at lowest y point, move to highest y
              point.
118.       for(int y = (int)minY;y<maxY;y = y + SQUARE_SIZE-1) {
```

```
119.            Point2D.Double intersect = new Point2D.Double();
120.            Line2D.Double vertLine = new Line2D.Double(minX,
                    y, maxX, y);
121.            boolean isIntersect = getIntersection(vertLine,
        line, intersect);
122.            if(isIntersect) {
123.                int gridX = ((int)intersect.x/SQUARE_SIZE);
124.                int gridY = ((int)intersect.y/SQUARE_SIZE);
125.                if(endSquares.getValue(gridX, gridY) == 0)
126.                    endSquares.changeValue(gridX, gridY,
                        (byte)+1);
127.            }
128.        }
129.      return endSquares;
130.    }
131.
132.    static private boolean trimVertical(Point2D.Double
            point1, Point2D.Double point2, Point2D.Double point3,
            Line2D.Double vertical) {
133.        Line2D.Double line1 = new Line2D.Double(point1.x,
                point1.y, point2.x, point2.y);
134.        Line2D.Double line2 = new Line2D.Double(point3.x,
                point3.y, point2.x, point2.y);
135.        Line2D.Double line3 = new Line2D.Double(point1.x,
                point1.y, point3.x, point3.y);
136.        Line2D.Double [] lines = {line1, line2, line3};
137.
138.        Point2D.Double [] intersects = new Point2D.
            Double[2];
139.
140.        int intersectCount = 0;
141.
142.        for(int i=0;i<3;i++) {
143.            intersects[intersectCount] = new Point2D.
                Double();
144.            boolean isIntersect = getIntersection(vertical,
                lines[i], intersects[intersectCount]);
145.            if(isIntersect) {
146.                ++intersectCount;
147.            }
148.            if(intersectCount >=2) break;
149.        }
150.
151.        if(intersectCount < 2)
152.            return false;
153.        else {
154.            vertical.x1 = intersects[0].x;
155.            vertical.x2 = intersects[1].x;
156.            vertical.y1 = intersects[0].y;
157.            vertical.y2 = intersects[1].y;
```

```
158.            return true;
159.        }
160.    }
161.
162.    static private Point2D.Double getPoint(int length,
            double angle) {
163.        double x = length * Math.cos(Math.toRadians(angle));
164.        double y = length * Math.sin(Math.toRadians(angle));
165.        return new Point2D.Double(x, y);
166.    }
167.
168.    static boolean getIntersection(Line2D.Double l1,
        Line2D.Double l2,
169.            Point2D.Double intersection) {
170.        if (!l1.intersectsLine(l2))
171.            return false;
172.
173.        double  x1 = l1.getX1(), y1 = l1.getY1(),
174.        x2 = l1.getX2(), y2 = l1.getY2(),
175.        x3 = l2.getX1(), y3 = l2.getY1(),
176.        x4 = l2.getX2(), y4 = l2.getY2();
177.
178.        intersection.x = det(det(x1, y1, x2, y2), x1 - x2,
179.                det(x3, y3, x4, y4), x3 - x4)/
180.                det(x1 - x2, y1 - y2, x3 - x4, y3 - y4);
181.        intersection.y = det(det(x1, y1, x2, y2), y1 - y2,
182.                det(x3, y3, x4, y4), y3 - y4)/
183.                det(x1 - x2, y1 - y2, x3 - x4, y3 - y4);
184.
185.        return true;
186.    }
187.
188.    static double det(double a, double b, double c,
            double d) {
189.        return a * d - b * c;
190.    }
191. }
```

In the previous chapter we noted that the robot will drift from the true coordinates after some movements. To remedy this, the robot automatically stops and asks to be placed at the starting position after a certain number of moves. You will have to manually pick up the robot and place it back at the starting position, effectively recalibrating the location. However, this will limit the distance the robot can explore.

```
1. import icommand.nxt.*;
2. import icommand.navigation.*;
3. import java.awt.geom.Point2D;
4.
5. public class Magellan {
```

```
6.
7.    static private int SQUARE_SIZE = 10;
8.    static private int SCAN_DEGREES = 30;
9.    static private int SCANS = 360/SCAN_DEGREES;
10.   private MapData md = new MapData();
11.   private MapDisplay display = new MapDisplay(md,
         SQUARE_SIZE);
12.   UltrasonicSensor us;
13.   Navigator robot;
14.
15.   public Magellan() {
16.       robot = new TachoNavigator(5.6F, 15F, Motor.B,
             Motor.C);
17.       us = new UltrasonicSensor(SensorPort.S3);
18.       Motor.A.setSpeed(400); // Scanner
19.   }
20.
21.   public double[][] scanCircle() {
22.       double [][] directions = new double[SCANS][2];
23.       // Rotate scanner back
24.       int starting = -180 + (SCAN_DEGREES/2);
25.       // Start loop
26.       for(int i=0;i<SCANS;i++) {
27.           Motor.A.rotateTo(starting + (SCAN_DEGREES * i));
28.           Point2D.Double pos = new Point2D.Double(robot.
             getX(), robot.getY()); // TEMP, will get from Nav
29.           double robotAngle = robot.getAngle();
30.           int sensorAngle = Motor.A.getTacho() + 180;
31.           double direction = sensorAngle + robotAngle;
32.           if(direction < 0) direction += 360;
33.           if(direction >= 360) direction -= 360;
34.           int distance = us.getDistance();
35.           directions[i][0] = direction;
36.           directions[i][1] = distance;
37.           MapScanner.plotSquares(pos, direction,
                 distance, md);
38.           display.update();
39.       }
40.       // Rotate scanner forward
41.       Motor.A.rotateTo(0);
42.       return directions;
43.   }
44.
45.   public void resetDrift() {
46.       Sound.playTone(1000, 1000);
47.       try {
48.           Thread.sleep(20000);
49.       } catch (Exception e) {}
50.   }
51.
```

```
52.     public static int chooseDirection(double[][] dirs) {
53.         int choice;
54.         do {
55.             choice = (int)(Math.random() * dirs.length);
56.         } while(dirs[choice][1] < 50);
57.         return choice;
58.     }
59.
60.     public void gotoLocation(Point2D.Double dest) {
61.         robot.gotoPoint(dest.x, dest.y);
62.     }
63.
64.     public static Point2D.Double getCoordinates(double
            [] dest) {
65.         Point2D.Double p = new Point2D.Double();
66.         if (dest[1] > 180) dest[1] = 180;
67.         p.x = Math.cos(Math.toRadians(dest[0])) * (dest[1]
                - 10);
68.         p.y = Math.sin(Math.toRadians(dest[0])) * (dest[1]
                - 10);
69.         return p;
70.     }
71.
72.     public static void main(String [] vals) {
73.         int DRIFT_RESET = 5;
74.         icommand.nxt.comm.NXTCommand.open();
75.
76.         Magellan mag = new Magellan();
77.
78.         for(int i=0;i<10;i++) {
79.             for(int j=0;j<DRIFT_RESET;j++) {
80.                 double [][] directions = mag.scanCircle();
81.                 // Pick one
82.                 int pick = Magellan.chooseDirection
                        (directions);
83.                 // Calculate coordinates and go
84.                 Point2D.Double dest = getCoordinates
                        (directions[pick]);
85.                 mag.gotoLocation(dest);
86.             }
87.             mag.resetDrift();
88.         }
89.         icommand.nxt.comm.NXTCommand.close();
90.     }
91. }
```

Using the Mapping Robot

You can substitute the TachoNavigator for a CompassNavigator if you own a compass sensor. The DRIFT_RESET constant determines the number of scans Magellan performs before it needs to be reset to the original position. If you are using CompassNavigator, then 10 is a good number, otherwise use a number closer to 5. It's a judgment call, depending on the floor surface. After this number of scans it will beep and wait for 20 seconds before starting again, so in this 20 second interval you must move it back to the origin point. Don't forget to mark the origin point and direction with a few coins.

Clear the area of objects like cords and blankets that might trip up the robot. Solid box-like objects like a couch or cabinet are fine. Now give it a try. The robot will draw a crude map of the area.

TRY IT: *Try attaching the light sensor to your robot and altering your code to collect a second set of map data consisting of light intensity readings. You will need to alter the mapping software to produce brighter colors for high numbers and darker colors for low numbers. When it is done you will probably be able to identify lamps or windows on the map, depending on the time of day.*

Path-finding

Topics in this Chapter

- Path-finding theory
- Blind Path-finding
- Mission Planning

Chapter 14

The mapping robot in the previous chapter used a strategy to move around the room and explore locations. Although this may seem like path-finding, it is actually just an exploration algorithm. Path-finding is the ability to avoid obstacles while moving from one specific point to another.

So far we have covered localization (where am I?) and mapping (where have I been?), and now it is time to cover path planning: How do I get there?

Pathfinding

Mobile robots need to get from one point to another, but sometimes the path is blocked; the robot can't travel a straight line from point A to point B (see Figure 14-1). Instead, the robot must find a path around obstacles to arrive at the destination.

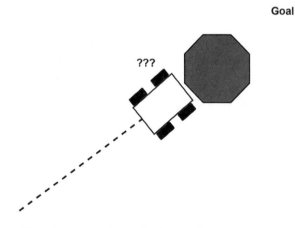

Figure 14-1 The robot encounters a blocked path

There are many strategies for path-finding. The best strategy often depends on the environment a robot must overcome. For example, if the robot is simply moving from one point to another across an area with many small obstacles, a blind path-finding strategy might work best. If the area contains many walls and rooms, then the blind path-finding strategy will fail because a decision might lead to a dead end. In this case, a map of the area will help the robot plan the best path. Let's examine these strategies.

Blind Path-finding

There are two competing philosophies in artificial intelligence. The first approach is traditional symbolic processing, which uses a high degree of computation to achieve a goal. The second approach, introduced by Professor Rodney Brooks of MIT, prefers simple biologically inspired solutions to complete tasks. He presented this theory in his paper, *Elephants Don't Play Chess*.

A blind path-finding strategy incorporates the Brooks philosophy. The robot simply travels straight at the destination. If it hits an obstacle, it travels one way or the other along the edge of the obstacle until it meets up with the original path (see Figure 14-2). No map data is required and the robot doesn't need to remember where it has been.

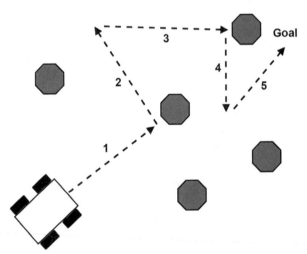

Figure 14-2 Blind path-finding

Of course, there are advantages and disadvantages to this strategy. Since it doesn't keep track of where it has been, nor does it plan ahead, the robot may become stuck, or not find the optimal path to the destination. The advantages are that it can deal with surprises like moving objects and it is easy to program.

Programming the Blind Path-finder

The algorithm is simple—start driving to the destination. If the ultrasonic sensor detects an object in the path, stop and look left and right. Drive in whichever direction has a clearer path (a longer ultrasonic distance). From the clear path, try driving to the destination again. We'll use Magellan from the previous chapter to try this since he has an ultrasonic sensor that rotates.

```
1.  import lejos.nxt.*;
2.  import lejos.navigation.*;
3.
4.  public class Pathfinder {
5.
6.     Navigator navigator;
7.     UltrasonicSensor us;
8.
9.     public Pathfinder(Navigator nav) {
10.       navigator = nav;
11.       us = new UltrasonicSensor(SensorPort.S3);
12.    }
13.
14.    public void goTo(float x, float y) {
15.
16.       navigator.goTo(x, y, true); // Start moving towards
              goal
17.
18.       // If detects object < distance remaining, check 30
              degrees left/right, use longest distance path
19.       while(navigator.isMoving()) {
20.         try {Thread.sleep(100);} catch(Exception e) {}
21.         int distance = us.getDistance();
22.         LCD.drawInt(distance, 3, 3);
23.         LCD.refresh();
24.         if(distance  < 40) {
25.           navigator.stop();
26.           Motor.A.rotateTo(-90);
27.           int leftD = us.getDistance();
28.           Motor.A.rotateTo(90);
29.           int rightD = us.getDistance();
30.           Motor.A.rotateTo(0);
31.           int jumpD = rightD>leftD?rightD:leftD;
32.           if(jumpD > 50) jumpD = 50;
33.           if(rightD > leftD) navigator.rotateTo(-90);
34.           else navigator.rotateTo(90);
35.           navigator.travel(jumpD);
36.
37.           this.goTo(x, y); // Recursive
38.         }
39.       }
40.    }
```

```
41.
42.   public static void main(String [] args) throws
         Exception {
43.      Navigator magellan = new TachoNavigator(5.6F, 14.0F,
            Motor.B, Motor.C, true);
44.      magellan.setSpeed(600);
45.      Pathfinder pf = new Pathfinder(magellan);
46.      pf.goTo(300,0);
47.   }
48. }
```

The code above begins heading for a point 300 centimeters directly ahead. The method seek() begins driving and watches for objects. If it encounters an object, it checks both ways and goes to the larger clear space. It then calls itself recursively to continue on its way.

This algorithm is not foolproof but it does a good job of finding its way through small obstacles in an open area. It can become stuck if placed in a room with walls, where the target is behind one of the walls (see Figure 14-3), however it does demonstrate the basic concept of blind path-finding.

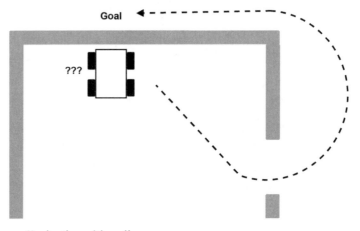

Figure 14-3 Navigation with walls

Mission Planning

In chapter 12 we looked at four aspects of navigation. The last one was mission planning, which asks the question, where am I going? Sometimes the answer is very simple to determine, for example a static goal like a recharging station that is always in the same place.

Sometimes the goal changes constantly. If the robot is given the task of scoring a goal in soccer, it must navigate between other players to get to the goal. To make a shot it needs to be in an unoccupied area in front of the net. The ball itself must navigate into the net without hitting the goaltender. A task like this requires detailed mission planning that can react to changes on the field.

Mission planning is in many ways the most interesting part of navigation. All the navigation topics (localization, mapping, path-finding) are merely functions that help accomplish this goal. Once these basic functions are mastered, your robot can move on to complex missions.

Think of your robot as a new baby, its memory clear. It takes a baby time to master the fundamentals of movement (how to crawl, how to walk) and it can't perform complex tasks. It can only start to perform interesting acts once it has learned the basics of movement.

Intelligent behavior arises with mission planning. There are no strict limits for mission planning. You can give your robot any mission, from getting a drink from the fridge to finding clothes on the floor and taking them to a hamper.

It was difficult to create complex mission planning with the RCX. By the time localization was programmed you were almost out of memory. It is my hope that during the life of NXT we will master localization, mapping, and path-finding, allowing programmers to concentrate more on mission planning.

Hands & Exoskeletons

Topics in this Chapter

- Robot Grippers
- Exoskeletons
- Data Glove Control

Chapter 15

R obotics seeks to imitate many of the functions humans perform every day. So far this book has looked at some ways to artificially create the major functions of the human body. This chapter will look at hands, one of the defining features of humans and primates.

It would be nice to attach a hand to a three-axis robot arm, but this is not possible with the parts in the NXT kit because of the limited number of motors. Instead, we will use one of the expansion solutions found in Chapter 19 to add an extra RCX motor to a robot.

If you don't have the extra parts needed to add a motor, you can still explore robot hands with the second project which explores the concept of artificially powered exoskeletons, such as those seen in *Aliens* (1986). The final project explores the exciting concept of robot control using a data glove.

A Robot Gripper

Our hands are unique in the animal world. Hands allow us to express our intelligence through engineering. There are other intelligent animals on the planet, and even animals with larger brains than humans, but without hands they have no way to express their intelligence. Dolphins may be smart, but without hands they can't create underwater kingdoms.

Many animals use a simple claw to clamp down on objects. This is what we will attempt to create in this section. Since we are all out of motors, we need some extra parts in order to add a gripping claw. For this project, we will use only parts from the Robotics Invention System, which many readers already own.

There is also the problem of controlling this motor, since the NXT only has three motor ports. To get around this limitation we will use the Mindsensors.com infrared link (described in more detail in chapter 19).

NOTE: *There are other options to increase the number of motors. See Chapter 19 for more details.*

Building the Gripper

It's time to pull your Robotics Invention System kit out of storage. You will require the RCX unit (with 6 AA batteries), one RCX motor, and some LEGO parts.

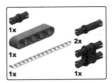

STEP 1 Attach the 5-unit beam using a pin and a long pin, as shown. Next plug in the long pin and double pin as shown.

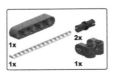

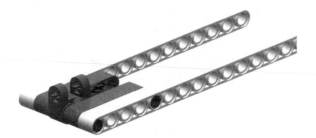

STEP 2 Insert the split-perpendicular axle connector as shown, then add the 5-unit beam and the long beam.

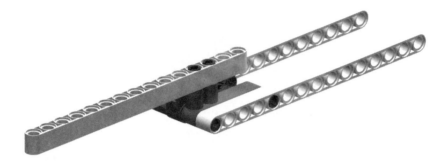

STEP 3 Insert the 2-unit axle into the split-perpendicular axle connector, as shown. Attach the remaining parts.

STEP 4 Add parts as shown.

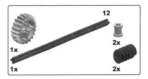

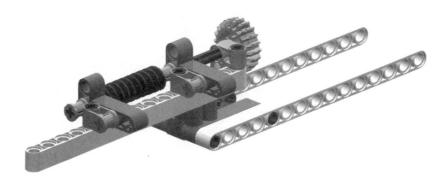

STEP 5 Make sure the worm gears are aligned to form one continuous work gear.

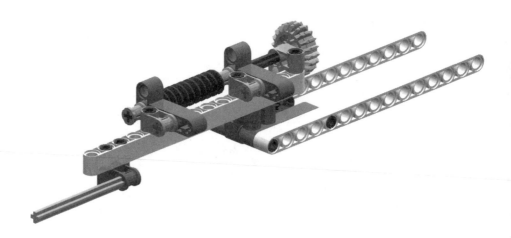

STEP 6 Add parts as shown.

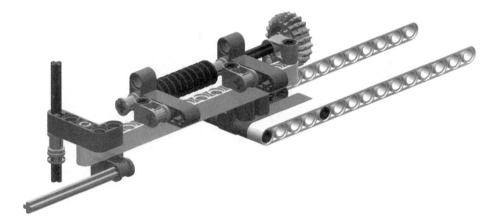

STEP 7 Add parts as shown.

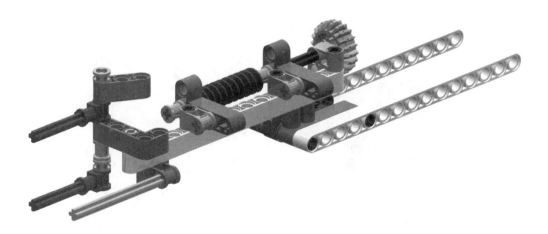

STEP 8 Add parts as shown.

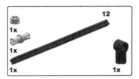

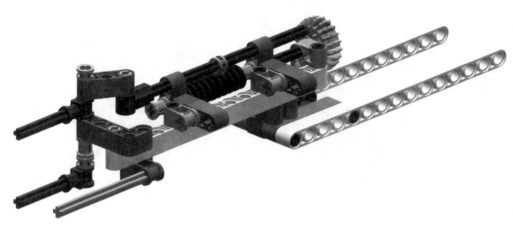

STEP 9 Add parts as shown.

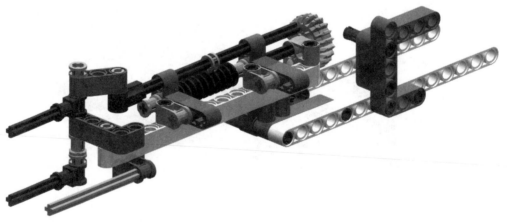

STEP 10 Add parts as shown.

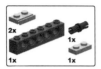

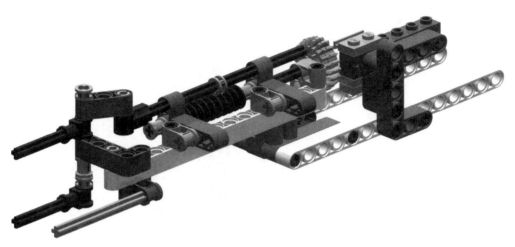

STEP 11 Now comes the RIS kit parts. The studded brick attaches to the L-beam using an additional pin.

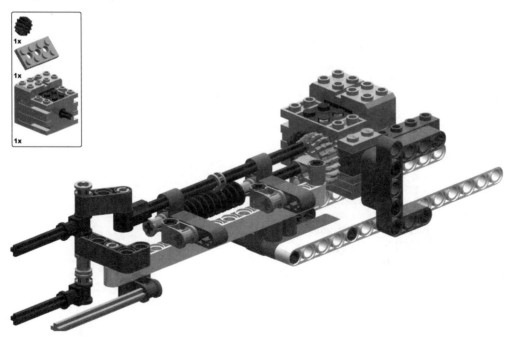

STEP 12 Slide the RCX motor on the supports and attach a plate to the bottom of the motor to secure it. Add the gear to the motor shaft. Read on for instructions on how to attach it to the arm.

The gripper attaches to Futura, the robot arm from Chapter 8. Remove the forearm of the regular model and replace it with the gripper (see Figure 15-1).

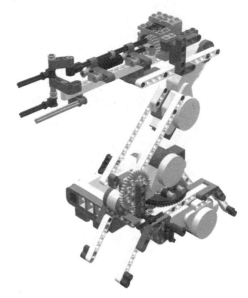

Figure 15-1 Attaching the gripper to Futura

Coding the Claw

To control the RCX brick (and the claw) we need to use the RCXLink class. We will modify the Arm.java code from chapter 8. Add the following object to the Arm class:

```
private RCXLink rcx;
```

Next add the following line to the Arm constructor:

```
rcx = new RCXLink(SensorPort.S1);
```

Now we need methods to open and close the claw:

```
1. public void openClaw() {
2.     rcx.A.forward();
3.     try {Thread.sleep(2500);}catch(Exception e){}
4.     rcx.A.stop();
5. }
6. public void closeClaw() {
7.     rcx.A.backward();
8.     try {Thread.sleep(2500);}catch(Exception e){}
9.     rcx.A.stop();
10. }
```

You can now use the modified arm code from Chapter 8 to control the new claw. Instead of hooking LEGO parts, try picking them up with the claw.

Exoskeletons

In this chapter we will create an exoskeleton. There have been numerous fictional portrayals of exoskeletons, most famously in the movie *Aliens* (1986), where Ripley dons an exoskeleton suit to battle an Alien queen. Robert Heinlein also described exoskeleton suits that allowed soldiers to jump superhuman heights in his novel *Starship Troopers*.

An artificially powered exoskeleton attaches directly to the user (or perhaps the user attaches to the exoskeleton) and then the exoskeleton translates normal movements into enhanced movements that might be faster, more powerful, or in other ways extend normal human capabilities.

Our task in this project is to build a human hand exoskeleton. Compared to arms, hands are very complicated. The hand has 24 bones and 16 joints (see Figure 15-2). The joints at the base of your fingers can move on multiple axes, bringing the total to 21 axes of rotation. If you used a motor for every joint, a robot hand would need 21 motors. That's not going to happen with the basic NXT parts.

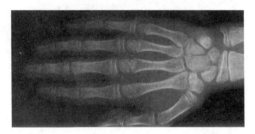

Figure 15-2 The human hand

Out of all the projects in this book, this is the only one that connects to you. It won't give you super strength, but it will increase the length of your arm.

Building the Exoskeleton

I was tempted to use the P5 glove (below) for this project, since it is ideal for sensing hand movements, but decided against it because of two problems. First, not enough readers own one. Second, the P5 cable to the glove would limit the mobility of the exoskeleton. Instead, I stuck to 100% NXT parts so everyone could try this project.

The goals for this project are as follows:

- Simulate a human hand with four fingers and a thumb
- Extend the length and size of the users regular hand (Did I mention the hand should be really big?)
- Allow direct user control of the exoskeleton via hand movement

How are we going to replicate the 21 articulated joints in a human hand? We only have three independent motors. Obviously we will have to make some compromises. First, we will use one motor to make all the fingers move at the same time. Second, we will reduce each finger to two joints instead of three. This reduces the number of gears we will need.

The next problem is control of the hand. We will use the rotation sensor inside an NXT motor. As you move your hand, it will turn the NXT motor, which can be translated to movements of the robot hand. We'll call our robot hand the T-800H, because of its similarity to the skeletal limbs of the Terminator.

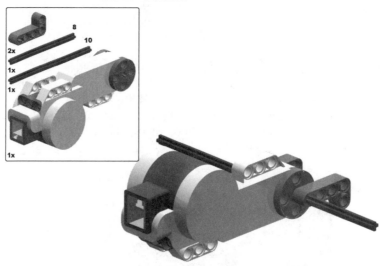

STEP 1 Insert the axles into the motor. The axles meet in the middle of the motor.

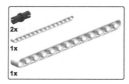

STEP 2 Add parts as shown.

STEP 3 Add parts as shown.

STEP 4 Add parts as shown.

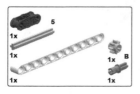

STEP 5 Add parts as shown.

STEP 6 Add parts as shown.

STEP 7 Add parts as shown.

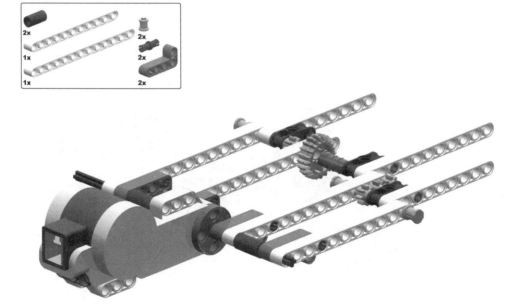

STEP 8 Add parts as shown.

STEP 9 Attach one 8-tooth gear on a yellow axle-pin. Then insert the 4-unit axle and attach the other 8-tooth gear. This attaches to the finger on the other side, as shown.

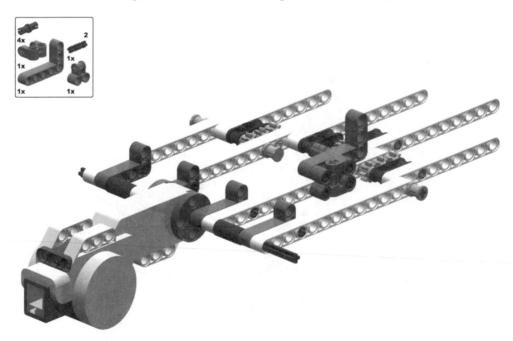

STEP 10 Add parts as shown.

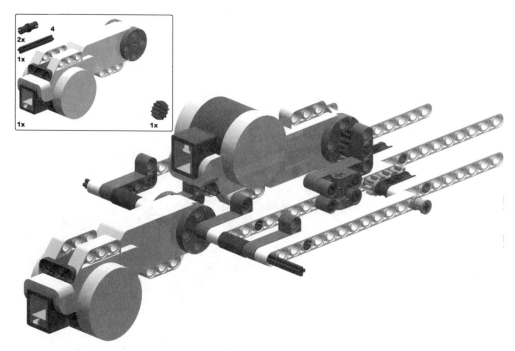

STEP 11 The motor attaches to the small L-beams using two pins, as shown. Insert the 4-unit axle into the motor and attach the black double-bevel gear.

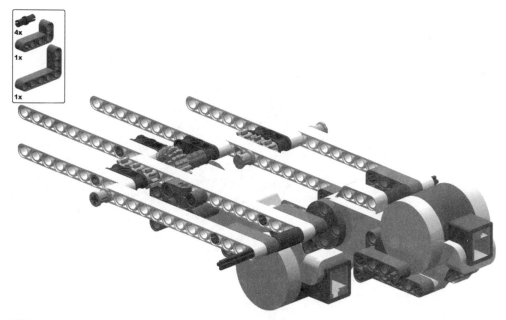

STEP 12 Turn the hand around. Attach the large L-beam to the motor using two pins. Next, attach the small L-beam to the large L-beam using two pins.

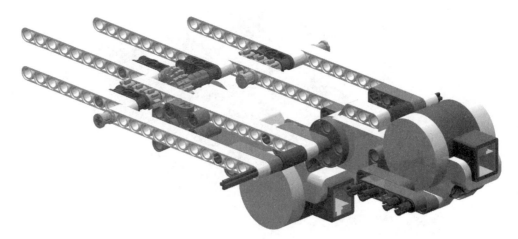

STEP 13 Insert two long pins and the blue axle-pin into the remaining holes on the small L-beam. Attach the 5-unit beam and add two pins to the end, as shown.

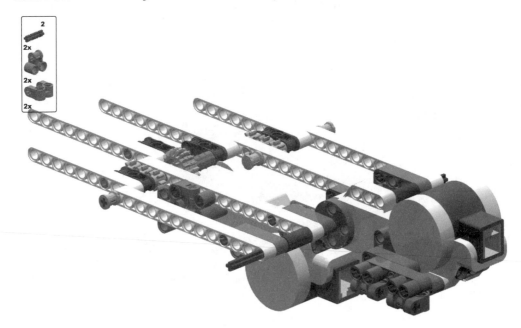

STEP 14 Add parts as shown.

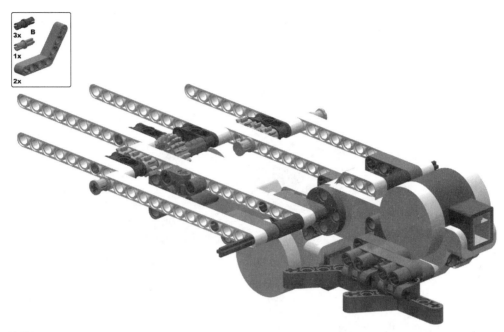

STEP 15 Insert the blue axle-pin into the bent beam, and right above that hole place a black pin. Attach this beam to the upper part as shown. Now attach the other bent beam using two black pins.

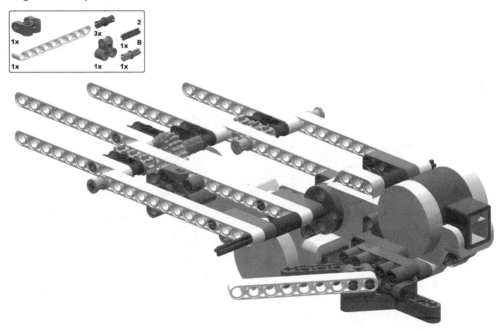

STEP 16 Insert a blue axle-pin and a black pin into the lower bent-beam from the previous step. Now attach the joiner. The beam attaches to the joiner with two pins, as shown.

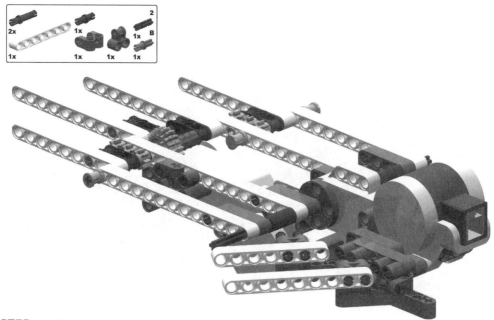

STEP 17 This joiner attaches to the bent beam in the same manner as the previous step. Use long pins to attach the white beam.

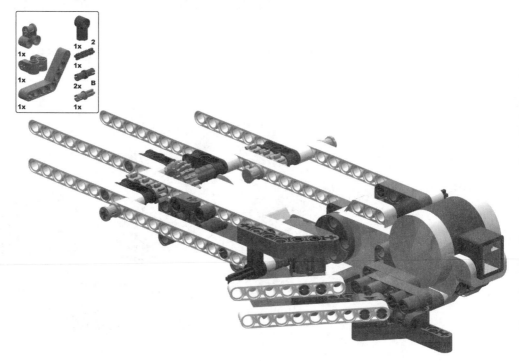

STEP 18 The joiner attaches to the long pins from the previous step. Place the black angle-connector on the axle as shown. Attach the beam to the joiner and connector as shown, using a blue axle-pin.

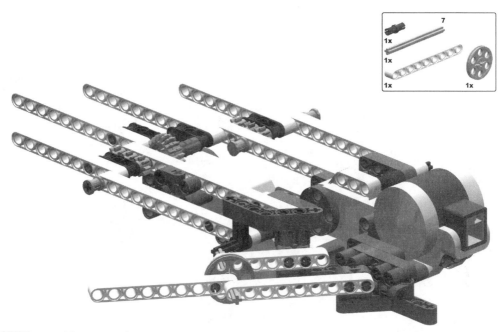

STEP 19 Add parts as shown.

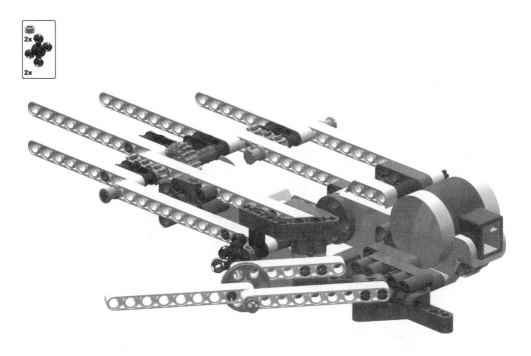

STEP 20 One half-bush attaches to the axle before the knob wheel. The other half-bush secures the thumb axle.

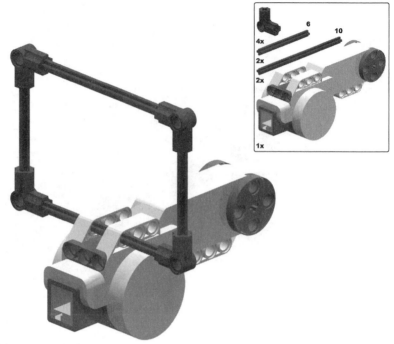

STEP 21 Add parts as shown.

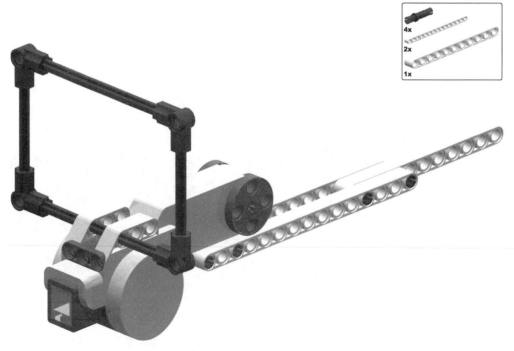

STEP 22 Add parts as shown.

STEP 23 Add parts as shown.

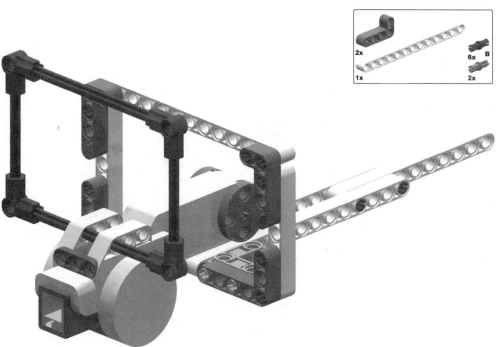

STEP 24 Add parts as shown.

STEP 25 Add parts as shown.

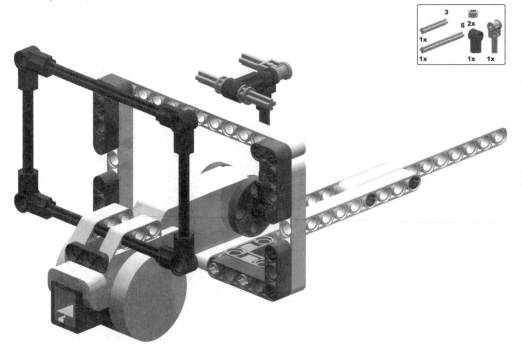

STEP 26 Add parts as shown.

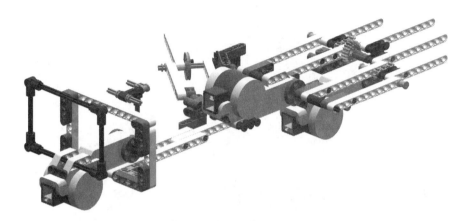

STEP 27 Attach the hand to the main unit using three long-pins with bushes.

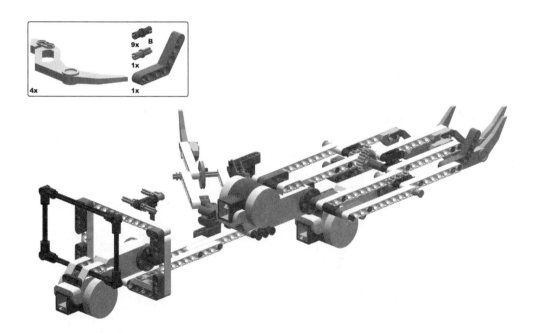

STEP 28 Add the claws to the ends of the fingers, as shown.

Using the T-800H

Before putting on the exoskeleton, attach the NXT brick to your wrist using rubber bands (see Figure 15-3). Once that is in place, fit your hand through the wrist hoops. Place the finger control between your first two fingers so that when you bend your fingers the motor rotates. Now plug in the wires – port A uses the short cable, port B uses a medium cable and port C uses a long cable. As you might guess, anything made from hard plastic is not very comfortable, but it works quite well, considering this is LEGO.

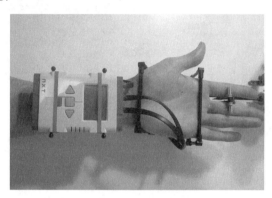

Figure 15-3 Inserting your hand into the T-800H

The hand is a little twitchy, but you can more or less make it perform as you want it to. It can't pick up heavy objects like a glass full of water, but it can pick up plastic cups, light clothing, and other light objects that are easy to grasp.

Controlling Robotics with a Data Glove

A *data glove* is an input device that fits on your hand and monitors the position of your fingers and hand. Usually these devices are used for virtual reality experiences, so they are sometimes called virtual reality gloves. This type of input device is used to control games on screen in a virtual world, but instead we will use it to control an arm in the real world.

The glove we will use for this project is the P5 glove, released in 2001 (see Figure 15-4A). Despite the commercial failure of this device, new unopened units are still widely available and can be found on eBay for as little as $10 (see Appendix A for purchasing options).

The P5 includes a tower unit that tracks the position and orientation of your hand (see Figure 15-4B). It does this by monitoring tiny infrared LEDs (light emitting diodes) on the glove itself. The glove also uses special flexible resistors to detect the position of each of your fingers. Let's explore how we can use this glove with Java.

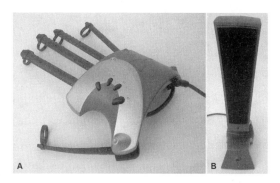

Figure 15-4 The P5 data glove and tower

Setting Up the P5 Glove

Before we can start programming the P5 glove there are some drivers that must be installed.

1. Install the P5 drivers using the setup program on the CD included with your P5 glove.

2. Download the open source drivers:
   http://tech.groups.yahoo.com/group/p5glove/
   Look for the file A_DualModeDriver****.zip (the version number might change as the software is updated).

3. Extract the files into a directory. e.g. c:\java\p5drivers

4. We need to build the Java classes. Go to a command prompt by selecting Start > Run and typing CMD [OK]

5. Change to the include directory.
   e.g. CD c:\java\p5drivers\include

6. Compile the code:
   javac -d . .\com\essentialreality\CP5DLL.java

7. Make a jar file by adding the com package to the jar.
   jar cvf CP5DLL.jar com

8. Add the location of the JAR file to your classpath setting.

9. If you are using Eclipse, add the JAR file to libraries.
   Select Project > Properties > Java Build Path > Libraries > Add External Jars... (see Figure 15-5). Browse to c:\java\p5drivers\include\CP5DLL.jar

10. Expand the CP5DLL.jar entry and select Native Library Location, Edit... and browse to C:/Java/P5drivers (see Figure 15-5).

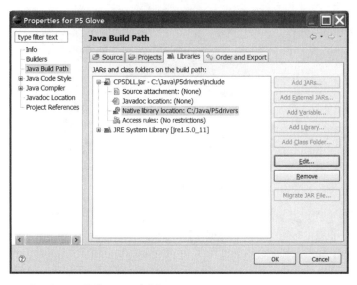

Figure 15-5 Setting up Eclipse variables

That's all! You are now ready to start programming your P5 glove.

Programming the P5 Glove

In this project, you will control a robot vehicle using hand gestures.
You can command the robot using the following gestures:

Command	Gesture
Stop	Flat hand
Forward	Point with index finger
Backward	Make a fist
Right	Roll hand to right
Left	Roll hand to left

Table 15-1 Hand gestures to control the robot

The lone class for communicating with the P5 glove is called CP5DLL.
Before doing anything, our code must first create a new instance of
CP5DLL and then initialize the glove by calling the P5_Init() method.

Once that is done, we can obtain information from the glove.
What kind? Well, lots, but you are probably interested in x, y, z, yaw,
pitch, roll, and the finger positions. These are all obtained from the
P5State nested class. The code below shows how to obtain the state
from CP5DLL.

The program can tell the position of each finger by looking at the finger[] array of P5State. If the value for a finger is close to 0, it means the finger is straight. If the finger is bent, the value will be as high as 63. The code below shows how all three hand gestures are identified.

```
1.  import icommand.navigation.Pilot;
2.  import icommand.nxt.*;
3.  import icommand.nxt.comm.*;
4.  import com.essentialreality.CP5DLL;
5.
6.  public class Merlin {
7.
8.      private CP5DLL gloves;
9.      private static CP5DLL.P5State gloveState;
10.     private int STOP = 0;
11.     private int FORWARD = 1;
12.     private int BACKWARD = 2;
13.     private int LEFT = 3;
14.     private int RIGHT = 4;
15.     Pilot robot = new Pilot(5.6F, 16.0F,Motor.B, Motor.C,
           true);
16.
17.     Merlin() {
18.       NXTCommand.open();
19.
20.       gloves = new CP5DLL();
21.       if (!gloves.P5_Init()) {
22.         System.out.println("Glove initialization failed");
23.         System.exit(1);
24.       }
25.       gloves.P5_SetForwardZ(-1); // Orient +ve Z towards
             user
26.       gloves.P5_SetMouseState(-1, false); // no mouse mode
27.       gloveState = gloves.state[0]; // get glove state
28.     }
29.
30.     public void run() {
31.       int oldState = 0;
32.       int state = 0;
33.       while(!gloveState.button[0]) { // Hit A to quit
34.         gloveState.update();
35.
36.         // 0=straight 63=fully bent
37.         int thumb = gloveState.finger[0];
38.         int index = gloveState.finger[1];
39.         int middle = gloveState.finger[2];
40.         int ring = gloveState.finger[3];
41.         int pinky = gloveState.finger[4];
42.
43.         // Identify hand gestures:
```

```
44.        if(index < 20 & middle > 40 & ring > 40)
45.           state = FORWARD; // Pointing
46.        if(index < 20 & middle < 20 & ring < 20)
47.           state = STOP; // Flat hand
48.        if(index > 40 & middle > 40 & ring > 40)
49.           state = BACKWARD; // Fist
50.
51.        float roll = gloveState.filterRoll;
52.        if(roll > 50)
53.           state = RIGHT;
54.        if(roll < -50)
55.           state = LEFT;
56.
57.        if(state != oldState) {
58.           if(state == FORWARD)
59.              robot.forward();
60.           else if(state == BACKWARD)
61.              robot.backward();
62.           else if(state == STOP)
63.              robot.stop();
64.           else if(state == LEFT)
65.              robot.rotate(360);
66.           else if(state == RIGHT)
67.              robot.rotate(-360);
68.
69.           oldState = state;
70.        }
71.     }
72.  }
73.
74.  public static void main(String[] args) {
75.     Merlin m = new Merlin();
76.     m.run();
77.     m.close();
78.  }
79.
80.  public void close() {
81.     gloves.P5_RestoreMouse(-1);
82.     gloves.P5_Close();
83.     NXTCommand.close();
84.  }
85. }
```

 NOTE: *The above code does not show how to obtain x, y, z coordinates or yaw and pitch. If you want to create a program that reads these values, use the following code:*

```
1. int x = gloveState.filterPos[0];
2. int y = gloveState.filterPos[1];
```

3. `int z = gloveState.filterPos[2];`
4. `int yaw = gloveState.filterYaw;`
5. `int pitch = gloveState.filterPitch;`

Results

This is a fun project. I named it Merlin because wizards are often depicted as being able to control objects using hand gestures. You can use any two wheeled robot in this book. I used Blighbot, the robot from chapter 12.

It's extremely easy and intuitive to control this robot using the hand gestures (see Figure 15-6 and 15-7). When you are finished, hit the A button on the glove to end the program.

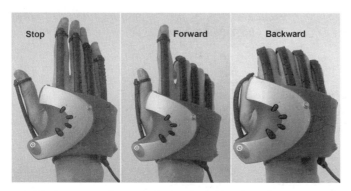

Figure 15-6 Hand gestures for stop, forward, and backward

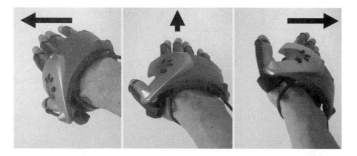

Figure 15-7 Hand gestures for steering

There is a slight, but barely noticeable, delay between giving a command and the robot's reaction. This is likely caused by the time it takes for the P5 drivers to decode the LED information, and the time it takes Bluetooth to send commands to the NXT brick.

TRY IT: *Try using the P5 glove to control one of the arm projects described earlier in this book. Since the arm code can position the hand at x, y, z coordinates and since the P5 glove can read x, y, z coordinates, it should be relatively easy to adapt the arm code for the P5.*

Network Robotics

Topics in this Chapter

- Communications API
- Telerobotics through the Internet

Chapter 16

T he Internet lets us bring back information from around the globe or send information to others. Have you ever wanted to do more than exchange information? How about something that allows you to reach right into your computer and physically manipulate something on the other side of the world? This chapter explores ways to send physical movement around the globe, right from your computer.

Communications API

The iCommand package has access to all the standard Java classes, including the communications API in java.io and java.net. The leJOS NXJ API also has these two packages, allowing data exchange from the NXT brick. The communications classes use streams, just like the standard java.io package, so anyone familiar with streams will find it easy to use.

The leJOS java.io package contains only the most basic streams relevant to sending and receiving data: InputStream, OutputStream, DataInputStream, and DataOutputStream. Input/Output Streams are the foundation of Streams, and they are useful only for sending bytes. If you want to send other data types such as characters, integers, and floating point numbers you will need to use data streams (see DataInputStream and DataOutputStream below).

java.io.InputStream

InputStream is the superclass representing input streams. Input streams transfer bytes. It is an abstract class so it can not be instantiated on its own. In leJOS NXJ, an instance of InputStream can be obtained from BTConnection (Bluetooth) or USBConnection by calling openInputStream().

- `public int read( ) throws IOException`

    Reads the next byte of data from the input stream. The value byte is returned as an int in the range 0 to 255. This method blocks (waits) until input data is available, the end of the stream is detected, or an exception is thrown.

- `public int read(byte[] b) throws IOException`

  Reads a number of bytes from the input stream and stores them into the buffer array b. The number of bytes actually read is returned as an integer. This method blocks until input data is available, end of file is detected, or an exception is thrown.

  Parameters
  b The buffer into which the data is read.

- `public int read(byte[] b, int off, int len) throws IOException`

  Reads up to len bytes of data from the input stream into an array of bytes. An attempt is made to read as many as len bytes, but a smaller number may be read, possibly zero. The number of bytes actually read is returned as an integer.

  Parameters
  b   The buffer into which the data is read.
  Off   The start offset in array b at which the data is written.
  Len   The maximum number of bytes to read.

- `public void close() throws IOException`

  Closes this input stream, calls flush() and releases any system resources associated with the stream.

java.io.OutputStream

OutputStream is the superclass of all classes representing an output stream of bytes. It is an abstract class so it can not be instantiated on its own. Its main function is to send a byte of data to a destination. Like InputStream, an instance of OutputStream can be obtained from BTConnection or USBConnection by calling openOutputStream().

- `public void write(int b) throws IOException`
  Writes the specified byte to this output stream.

- `public void write(byte b[]) throws IOException`

  Writes b.length bytes from the specified byte array to this output stream.

  Parameters
  b The data.

- `public void write(byte b[], int off, int len) throws IOException`

  Writes len bytes from the specified byte array starting at offset off to this output stream. The general contract for write(b, off, len) is that some of the bytes in the array b are written to the output stream in order; element b[off] is the first byte written and b[off+len-1] is the last byte written by this operation.

Parameters

b   The data.

off   The start offset in the data.

len   The maximum number of bytes to write.

- `public void flush() throws IOException`

    Flushes this output stream and forces any buffered output bytes to be written out. The general contract of flush() is that calling it is an indication that, if any bytes previously written have been buffered by the implementation of the output stream, such bytes should immediately be written to their intended destination.

WARNING: *Flush is one of the most important but often forgotten methods of streams. The non-use of this method probably accounts for most bugs when using the java.io package. Don't forget to call flush() after sending data, otherwise the data may never be sent to the destination!*

- `public void close() throws IOException`

    Closes this output stream and releases any system resources associated with this stream. A closed stream cannot perform output operations and cannot be reopened. A call to flush() is made just before the stream is closed.

java.io.DataInputStream

DataInputStream extends InputStream, so it has all the methods of InputStream (see above). This method allows data types other than bytes to be sent. This includes short, int, float, double, char, boolean and String. In order to save memory the leJOS DataInputStream class does not extend FilterInputStream nor does it implement a DataInput interface. You can obtain an instance of DataInputStream from BTConnection or USBConnection by calling openDataInputStream().

- `public DataInputStream(InputStream in)`

    Returns an instance of DataInputStream. The constructor requires an InputStream object.

    Parameters

    in   The input stream.

- `public final boolean readBoolean() throws IOException`

    Used to send a boolean value through a stream. Reads one input byte and returns true if that byte is nonzero, false if that byte is zero.

- `public final byte readByte() throws IOException`

    Reads and returns one input byte. The byte is treated as a signed value in the range -128 through 127, inclusive.

- `public final short readShort() throws IOException`

    Reads two input bytes and returns a short value.

- `public final char readChar() throws IOException`

    Reads an input char and returns the char value. (A Unicode char is made up of two bytes.)

- `public final int readInt() throws IOException`

    Reads four input bytes and returns an int value.

- `public final float readFloat() throws IOException`

    Reads four input bytes and returns a float value.

- `public final double readDouble() throws IOException`

    Reads eight input bytes and returns a double value.

java.io.DataOutputStream

If DataInputStream is the receiver then DataOutputStream is the sender. It encodes various data types into byte values and sends them across a data stream. DataOutputStream extends OutputStream, so it has all the methods described in the OutputStream API. DataOutputStream does not extend FilterOutputStream nor does it implement DataOutput. You can obtain an instance of DataOutputStream from either BTConnection or USBConnection by using openDataOutputStream().

- `public DataOutputStream(OutputStream out)`

    Creates a new data output stream to write data to the specified underlying output stream.

    Parameters
    `out`   The output stream.

- `public final void writeBoolean(boolean v) throws IOException`

    Writes a boolean value to this output stream.

    Parameters
    `v`   A boolean value.

- `public final void writeByte(int v) throws IOException`

    Writes to the output stream the eight low-order bits of the argument v.

    Parameters
    `v`   A byte value.

- `public final void writeShort(int v) throws IOException`

    Writes two bytes to the output stream to represent the value of the argument.

    Parameters
    `v`   A short value.

- `public final void writeChar(int v) throws IOException`

  Writes a char value, which is comprised of two bytes, to the output stream.

  Parameters
  v   A char value.

- `public final void writeInt(int v) throws IOException`

  Writes an int value, which is comprised of four bytes, to the output stream.

  Parameters
  v   An int value.

- `public final void writeFloat(float v) throws IOException`

  Writes a float value, which is comprised of four bytes, to the output stream.

  Parameters
  v   A float value.

- `public final void writeDouble(double v) throws IOException`

  Writes a double value, which is comprised of eight bytes, to the output stream.

  Parameters
  v   A double value.

javax.microedition.io.StreamConnection

StreamConnection is an interface at the top of the communications hierarchy. On the leJOS project we are trying to emulate the Java APIs that are made for portable devices, so we decided to use javax.microedition.io. There are two StreamConnection classes in leJOS NXJ: BTConnection and USBConnection. StreamConnection objects function much like java.net.Socket in standard Java. These classes have in common the ability to allocate InputStream and OutputStream objects, which are vital for sending and receiving data in Java.

- `public InputStream openInputStream()`

  Returns an input stream for thisStreamConnection.

- `public OutputStream openOutputStream()`

  Returns an output stream for this StreamConnection.

- `public DataInputStream openDataInputStream()`

  Returns a data input stream for thisStreamConnection.

- `public DataOutputStream openDataOutputStream()`

  Returns a data output stream for this StreamConnection.

- `public void close()`

  Closes this DataPort.

BTConnection and USBConnection

The classes BTConnection and USBConnection both implement StreamConnection. Both of these classes are in lejos.nxt.comm. There is no need to go over their methods since they are identical to StreamConnection. However, it's not immediately obvious how to get one of these connections.

The answer lies in the Bluetooth and USB classes, which are also in the lejos.nxt.comm package. These classes have methods for sending and receiving data, but they are mostly used for obtaining a StreamConnection object. This is done using Bluetooth.waitForConnection(), as follows:

```
BTConnection btc = null;
btc = Bluetooth.waitForConnection();
```

That's all there is to it. The next two sections will present real examples using the Communications API.

Telerobotics with the Internet

The philosopher Marshall McLuhan described electronic media as an extension of our senses. For example, radio allows our ears to travel to a conversation at the top of Mount Everest. Television allows our eyes to see the bottom of the ocean. Electronic media extends the range of our senses.

McLuhan anticipated the Internet even before ARPANET. He saw it as a way for individuals to extend communications using their senses. McLuhan was very poorly understood in the fifties and sixties because no one understood what he was talking about. Today, with the Internet realized, McLuhan seems very straightforward.

Using the LEGO NXT, we can take McLuhan's concept of extending our senses even further. In this chapter, we will use the Internet (or any network) to explore telerobotics. If McLuhan were alive today, he would probably describe telerobotics as an extension of both our senses and our limbs. We will use your computer and network to control robots over any distance—in another room or another city. Let's make that happen.

Controlling an Arm through a Network

This project will allow you to control the robot arm Futura (Chapter 8) from anywhere across the globe. The architecture is a little more complex than the other projects in this book. It requires code to run on two separate platforms—a *client* computer and a *server* computer (Figure 16-1). Technically there is code running on three platforms, since the NXT brick also has code to handle commands. However, this code becomes active as soon as you turn on the brick. You don't have to worry about programming anything special for the NXT since the iCommand code does all the low level communications.

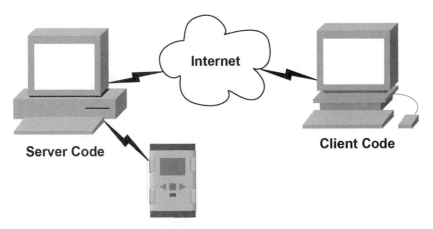

Figure 16-1 Architecture for robot control over the Internet

The client application will send user commands from the client to the server. The server acts as a sort of middleman, shuffling commands from the user to the NXT brick. Let's examine the server code first.

```
1.  import java.net.*;
2.  import java.io.*;
3.  import icommand.nxt.*;
4.
5.  public class TeleServer {
6.
7.     ServerSocket server;
8.     Socket client;
9.     DataOutputStream clientOutStream;
10.    DataInputStream clientInStream;
11.
12.    public TeleServer(int portNumber) {
13.       try {
14.          server = new ServerSocket(portNumber);
15.       }
16.       catch (IOException io){
17.          io.printStackTrace();
18.          System.exit(0);
19.       }
20.       icommand.nxt.comm.NXTCommand.open();
21.       Motor.C.setSpeed(150);
22.       Motor.B.setSpeed(500);
23.       Motor.A.setSpeed(600);
24.    }
25.
```

```
26.    public static void main(String [] args) {
27.      TeleServer host = new TeleServer(TeleClient.PORT);
28.      while(true) {
29.        host.waitForConnection();
30.        while(host.client!=null)
31.          host.waitForCommands();
32.      }
33.    }
34.
35.    /** Wait for a connection from the Client machine */
36.    public void waitForConnection() {
37.      try {
38.        System.out.println("Listening for client.");
39.        client = server.accept();
40.        clientOutStream = new DataOutputStream(client.
             getOutputStream());
41.        clientInStream = new DataInputStream(client.
             getInputStream());
42.        System.out.println("Client connected.");
43.      } catch(IOException io) {
44.        io.printStackTrace();
45.      }
46.    }
47.
48.    /** Wait for coordinates from the Client machine */
49.    public void waitForCommands() {
50.      int command = 0;
51.      try {
52.        command = clientInStream.readInt();
53.        System.out.println("Command received: " + command);
54.        if(command == TeleClient.CLOSE) {
55.          client.close();
56.          client = null;
57.          System.out.println("Client disconnected.");
58.        }
59.      } catch(IOException io) {
60.        System.out.println("Client unexpectedly
             terminated.");
61.        client = null;
62.      }
63.      System.out.println("Command " + command);
64.      switch (command) {
65.      case TeleClient.A_FORWARD:
66.        Motor.A.forward();
67.        break;
68.      case TeleClient.A_BACKWARD:
69.        Motor.A.backward();
70.        break;
71.      case TeleClient.A_STOP:
72.        Motor.A.stop();
```

```
73.        sendValue(Motor.A.getTacho());
74.        break;
75.      case TeleClient.B_FORWARD:
76.        Motor.B.forward();
77.        break;
78.      case TeleClient.B_BACKWARD:
79.        Motor.B.backward();
80.        break;
81.      case TeleClient.B_STOP:
82.        Motor.B.stop();
83.        sendValue(Motor.A.getTacho());
84.        break;
85.      case TeleClient.C_FORWARD:
86.        Motor.C.forward();
87.        break;
88.      case TeleClient.C_BACKWARD:
89.        Motor.C.backward();
90.        break;
91.      case TeleClient.C_STOP:
92.        Motor.C.stop();
93.        sendValue(Motor.A.getTacho());
94.        break;
95.      default:
96.        break;
97.      }
98.    }
99.
100.   private void sendValue(int value) {
101.     try {
102.       clientOutStream.writeInt(value);
103.     } catch(IOException io) {
104.       io.printStackTrace();
105.     }
106.   }
107. }
```

The next part of our project is the client code. It generates a friendly API for the user, allowing easy connection and control of the robot (see Figure 16-2). Let's examine this code:

```
1. import java.awt.*;
2. import java.awt.event.*;
3. import java.io.*;
4. import java.net.Socket;
5.
6. public class TeleClient extends Frame implements
   KeyListener{
7.   public static final int PORT = 7360;
8.   public static final int CLOSE = 0;
9.   public static final int B_FORWARD = 87, // W = main up
10.  B_BACKWARD = 83, // S = main down
```

```
11.    A_FORWARD = 65, // A = left
12.    A_BACKWARD = 68, // D = right
13.    C_FORWARD = 84, // T = forearm up
14.    C_BACKWARD = 71; // G = forearm down
15.
16.    public static final int A_STOP = 3;
17.    public static final int B_STOP = 6;
18.    public static final int C_STOP = 9;
19.
20.    Button btnConnect;
21.    TextField txtTachoA;
22.    TextField txtTachoB;
23.    TextField txtTachoC;
24.    TextField txtIPAddress;
25.    TextArea messages;
26.
27.    private Socket socket;
28.    private DataOutputStream outStream;
29.    private DataInputStream inStream;
30.
31.    public TeleClient(String title, String ip) {
32.      super(title);
33.      this.setSize(400, 300);
34.      this.addWindowListener(new WindowAdapter() {
35.        public void windowClosing(WindowEvent e) {
36.          System.out.println("Ending Telerobotics Client");
37.          disconnect();
38.          System.exit(0);
39.        }
40.      });
41.      buildGUI(ip);
42.      this.setVisible(true);
43.      btnConnect.addKeyListener(this);
44.    }
45.
46.    public static void main(String args[]) {
47.      System.out.println("Starting Telerobotics Client...");
48.      new TeleClient("Telerobotics Client", "127.0.0.1");
49.    }
50.
51.    public void buildGUI(String ip) {
52.      Panel mainPanel = new Panel (new BorderLayout());
53.      ControlListener cl = new ControlListener();
54.
55.      btnConnect = new Button("Connect");
56.      btnConnect.addActionListener(cl);
57.
58.      txtTachoA = new TextField("",20);
59.      //txtTachoA.setEditable(false);
60.      txtTachoB = new TextField("",20);
```

```
61.     //txtTachoB.setEditable(false);
62.     txtTachoC = new TextField("",20);
63.     //txtTachoC.setEditable(false);
64.     txtIPAddress = new TextField(ip,16);
65.
66.     messages = new TextArea("status: DISCONNECTED");
67.     messages.setEditable(false);
68.
69.     Panel north = new Panel(new FlowLayout(FlowLayout.
            LEFT));
70.     north.add(btnConnect);
71.     north.add(txtIPAddress);
72.
73.     Panel center = new Panel(new GridLayout(5,1));
74.     center.add(new Label("Use A-D, W-S, T-G to control
            robot"));
75.
76.     Panel center1 = new Panel(new FlowLayout(FlowLayout.
            LEFT));
77.     center1.add(new Label("A:"));
78.     center1.add(txtTachoA);
79.
80.     Panel center2 = new Panel(new FlowLayout(FlowLayout.
            LEFT));
81.     center2.add(new Label("B:"));
82.     center2.add(txtTachoB);
83.
84.     Panel center3 = new Panel(new FlowLayout(FlowLayout.
            LEFT));
85.     center3.add(new Label("C:"));
86.     center3.add(txtTachoC);
87.
88.     Panel center4 = new Panel(new FlowLayout(FlowLayout.
            LEFT));
89.     center4.add(messages);
90.
91.     center.add(center1);
92.     center.add(center2);
93.     center.add(center3);
94.     center.add(center4);
95.
96.     mainPanel.add(north, "North");
97.     mainPanel.add(center, "Center");
98.     this.add(mainPanel);
99.   }
100.
101.  private void sendCommand(int command){
102.    // Send coordinates to Server:
103.    messages.setText("status: SENDING command.");
104.    try {
```

```
105.        outStream.writeInt(command);
106.      } catch(IOException io) {
107.        messages.setText("status: ERROR Problems occurred
              sending data.");
108.      }
109.
110.      messages.setText("status: Command SENT.");
111.    }
112.
113.    /** A listener class for all the buttons of the GUI. */
114.    private class ControlListener implements
            ActionListener{
115.      public void actionPerformed(ActionEvent e) {
116.        String command = e.getActionCommand();
117.        if (command.equals("Connect")) {
118.          try {
119.            socket = new Socket(txtIPAddress.getText(),
                  PORT);
120.            outStream = new DataOutputStream(socket.
                  getOutputStream());
121.            inStream = new DataInputStream(socket.
                  getInputStream());
122.            messages.setText("status: CONNECTED");
123.            btnConnect.setLabel("Disconnect");
124.          } catch (Exception exc) {
125.            messages.setText("status: FAILURE Error
                  establishing connection with server.");
126.            System.out.println("Error: " + exc);
127.          }
128.        }
129.        else if (command.equals("Disconnect")) {
130.          disconnect();
131.        }
132.      }
133.    }
134.
135.    public void disconnect() {
136.      try {
137.        sendCommand(CLOSE);
138.        outStream.close();
139.        inStream.close();
140.        socket.close();
141.        btnConnect.setLabel("Connect");
142.        messages.setText("status: DISCONNECTED");
143.      } catch (Exception exc) {
144.        messages.setText("status: FAILURE Error closing
              connection with server.");
145.        System.out.println("Error: " + exc);
146.      }
147.    }
```

```
148.
149.    public int receiveValue() {
150.      int val = 0;
151.      try {
152.        messages.setText("status: WAITING for tachometer
                readings.");
153.        val = inStream.readInt();
154.        messages.setText("status: RECEIVED tachometer
                reading.");
155.      } catch(IOException io) {
156.        messages.setText("status: ERROR Data transfer of
                tacho failed.");
157.      }
158.      return val;
159.    }
160.
161.    public void keyPressed(KeyEvent e) {
162.      sendCommand(e.getKeyCode());
163.      System.out.println("Pressed " + e.getKeyCode());
164.    }
165.
166.    public void keyReleased(KeyEvent e) {
167.      switch(e.getKeyCode()) {
168.      case A_FORWARD:
169.      case A_BACKWARD:
170.        sendCommand(A_STOP);
171.        break;
172.      case B_FORWARD:
173.      case B_BACKWARD:
174.        sendCommand(B_STOP);
175.        break;
176.      case C_FORWARD:
177.      case C_BACKWARD:
178.        sendCommand(C_STOP);
179.        break;
180.      }
181.    }
182.
183.    public void keyTyped(KeyEvent arg0) {}
184. }
```

Since the client application is test code, it isn't as fleshed out as production code might be. Ideally, if someone was already connected to the server it would respond that the server is busy, with information about who was connected and how long they have been using the robot.

> **WARNING:** *The keyboard commands are attached to the Connect/Disconnect button, since that object is likely to hold the focus. If you click on other widgets in the window and the button loses focus, the keys stop working. Tab back to the button to regain focus.*

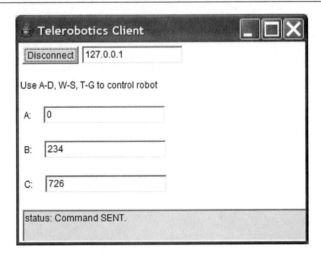

Figure 16-2 The Telerobotics GUI

Results

Now that we have all the code ready it's time to test it out. If you don't have two computers in your home, you can run both the client and the server on the same machine. First, run the server code on your PC. It will sit waiting for a client to connect.

Next, run the client (either on the same machine, or on a different machine on the network). Type in the IP address (or leave it as 127.0.0.1 if it is running on the same machine) and click connect.

You will need to know the IP address of the server computer in order to connect to it over a local network. Windows users can find this by going to a command prompt and typing: (see Figure 16-3)

```
ipconfig
```

Linux users can use a similar program in the super-user directory

```
/sbin/ifconfig
```

Mac OSX users can open system preferences. Under Internet and network, click network.

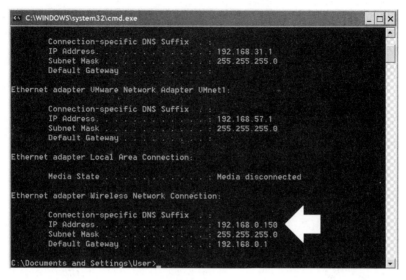

Figure 16-3 Finding your local IP Address

If you are running this project on a local network, all you need to know is listed above. If you want to connect to a server computer over the Internet, you will need to do something slightly different. Most networks use a router that connects to the Internet, allowing everyone to have Internet access. You need to know the IP address that the rest of the world sees. Sometimes this changes every time the router connects to the Internet provider, depending on your service. To check the IP address, open a browser and go to your router admin page (usually http://192.168.0.1). Find the status of your connection, which also lists the IP address (see Figure 16-4).

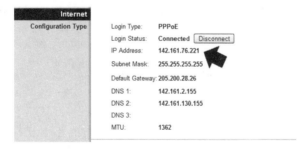

Figure 16-4 An example of the Internet IP address

You aren't finished yet. Since you probably have more than one computer on your router, you need to set up the router to forward packets to the server computer. It does this using port numbers. Every application uses a different port number, and our robot arm application uses port 7360, as you can see in the client code. You want any data meant for port 7360 to be forwarded to your server computer. You'll have to consult your documentation for how to set up *port forwarding* on your router, but it should look something like Figure 16-5.

Port Range					
Application	Start	End	Protocol	IP Address	Enable
leJOS	7360	to 7360	TCP	192.168.0.120	☑

Figure 16-5 Port forwarding

Now that you are ready to go, run the client application and click connect. You can now press the different keys that control the three different motors. After each press you will see the tachometer count that shows that the motors did in fact move.

NOTE: *If you are having problems accessing a server through the Internet, chances are there is a firewall blocking some ports. Temporarily disable the firewall while testing this project, or allow firewall access through port 7360.*

If you have a webcam, I recommend pointing it at Futura so you can experience telepresence. This will allow you do fun things with Futura, such as picking up rings with the hook. You can receive a live webcam image by using one of the many free *instant messaging* programs, such as Windows Live Messenger, Skype, AOL Instant Messenger (AIM), Yahoo! Messenger, or one of the many open source instant messaging clients built on Jabber.

TRY IT: *It's also possible to control the robot through a web page. Because this is a book on robotics and not web development, I won't go into these techniques in detail. Briefly, you'll need to set up your own web page server such as Apache or Resin. Java Server Pages (JSP) are ideal for this type of project. JSP can interact with beans, so you will need to code a simple bean that allows you to interact with some iCommand methods. Once the bean is in your server directory, your JSP can interact with the bean (and users can interact with it via a JSP page).*

Scanning

Topics in this Chapter

- 3D Object Scanning
- Copy Machine

Chapter 17

LEGO MINDSTORMS concentrates on robots, but you can also use it to build ordinary electronic devices. This chapter will focus on a couple of non-robotic devices.

The ultrasonic sensor is great for sensing the outside environment but it can also focus inward on one object. Although it can scan only the distance to an object, you can begin to construct a more complete picture by taking multiple measurements. The light sensor can be used in a similar manner. This chapter will explore two scanning methods.

3D Scanning

In Terminator 2, the T-800 robot enters a bar and scans various customers to see if their clothing will fit him. It's a great scene which visually shows the computations going on beneath T-800's human façade. This section will attempt to create a device that scans solid objects using only NXT parts.

The scanner in this section was inspired by the *Intellifit* body scanner, a device for taking clothing measurements (see Figure 17-1). This device scans a person and records their exact measurements. It was invented by former Commodore engineer Al Charpentier, who joked that it is the first part of a transporter device he's building.

Figure 17-1 The Intellifit body scanner (© 2007 Intellifit)

The Intellifit uses a long arm packed with shortwave sensors to detect the distance to a person (or any other water-based object). LEGO doesn't come with shortwave sensors, but the single ultrasonic sensor can slowly rotate around a stationary object and record data.

The ultrasonic sensor returns the closest part of an object within the detection cone (see Figure 7-2). This means the ultrasonic sensor is not as accurate as a laser (the dashed line in Figure 17-2), but often the closest point is good enough to give a reasonable representation of an object.

Figure 7-2 The cone effect when scanning an object

Building the Scanner

The first design for this scanner was like the Intellifit, in which the scanner rotates around a stationary object. This was quickly discarded in favor of a simpler design where the sensor remains stationary and the object rotates on a turntable. Let's try building the device.

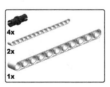

STEP 1 Add parts as shown.

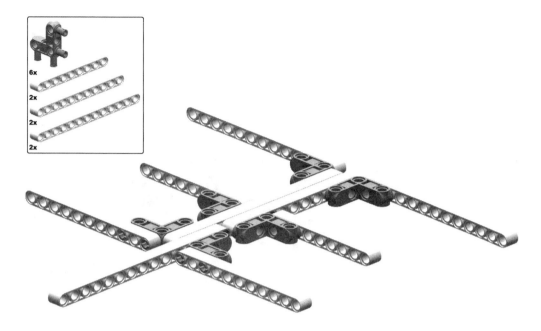

STEP 2 Add parts as shown.

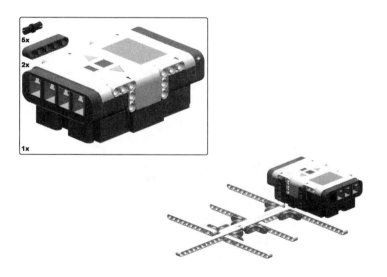

STEP 3 Attach one 5-unit beam to the NXT using two pins, and connect the other using only one pin. Use the remaining pins to attach the beams to the base.

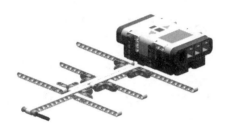

STEP 4 Add parts as shown.

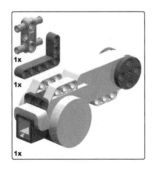

STEP 5 Add parts as shown.

STEP 6 Add parts as shown.

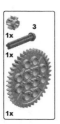

STEP 7 Add parts as shown.

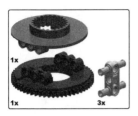

STEP 8 Insert a grey axle joiner on each side of the turntable. Attach to the motor using a third axle joiner.

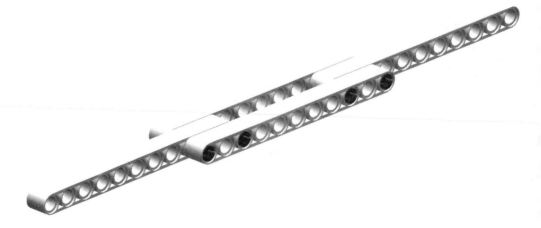

STEP 9 Add parts as shown.

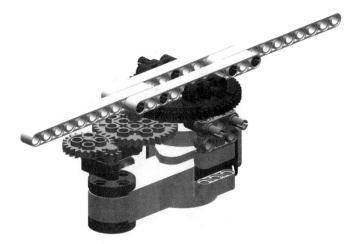

STEP 10 Add parts as shown.

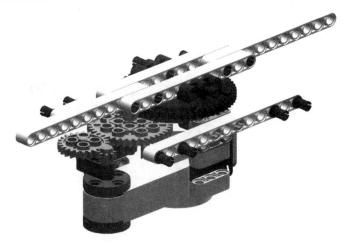

STEP 11 Attach beams on both sides of the turntable, as shown.

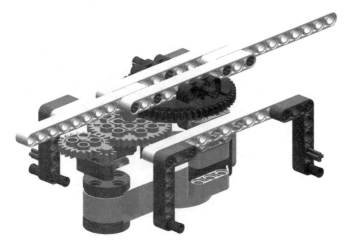

STEP 12 Attach a blue axle pin and black pin to the rear L-beams. The closest L-beam receives only one pin. The remaining L-beam receives a long pin and a blue axle pin.

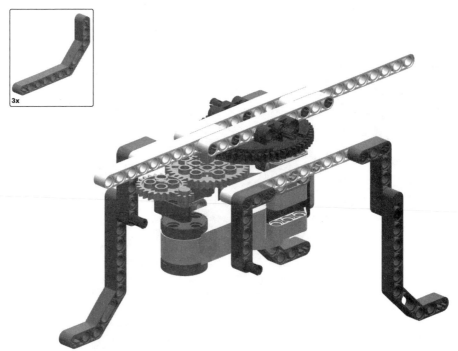

STEP 13 Add parts as shown.

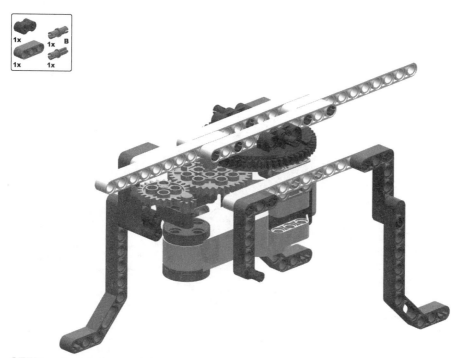

STEP 14 Attach the 3-unit beam using an additional black pin. Attach the rubber connector using a blue axle pin.

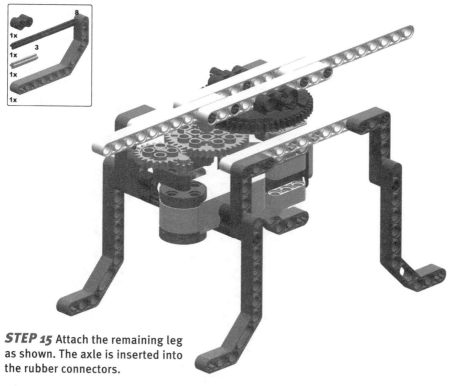

STEP 15 Attach the remaining leg as shown. The axle is inserted into the rubber connectors.

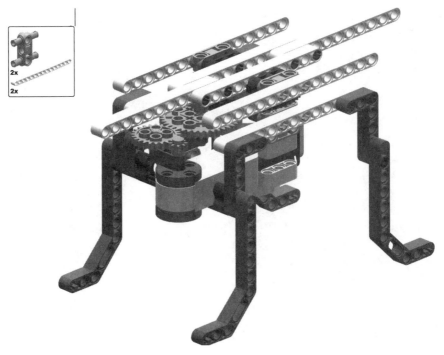

STEP 16 Add parts as shown.

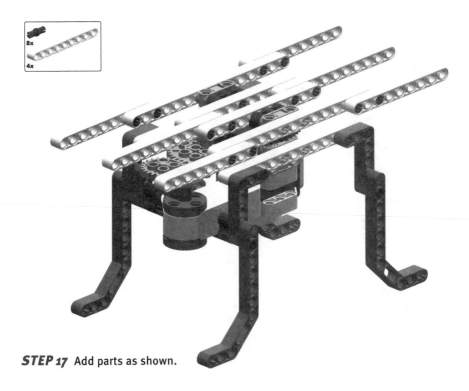

STEP 17 Add parts as shown.

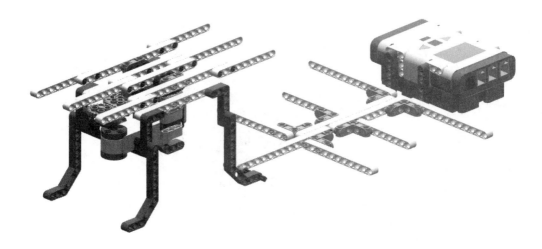

STEP 18 Attach the table to the base. This provides the proper spacing from the scanner unit.

STEP 19 Insert the axle so it is roughly in the middle of the tire. Add two bushes to the opposite side, and one on the close side.

STEP 20 Add parts as shown.

STEP 21 Attach the 5-unit beam to the L-beam using a single pin (obscured). Add the rest of the parts as shown.

STEP 22 Add parts as shown.

STEP 23 Add parts as shown.

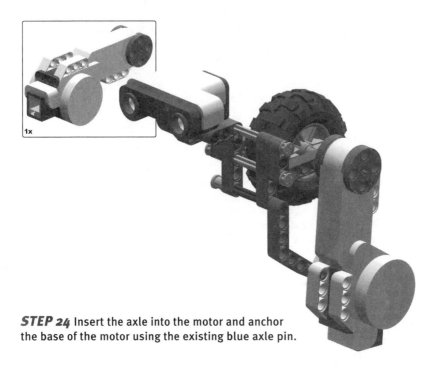

STEP 24 Insert the axle into the motor and anchor the base of the motor using the existing blue axle pin.

STEP 25 Add parts as shown.

STEP 26 Add parts as shown.

STEP 27 Start with the short beam, then begin attaching long beams as shown.

STEP 28 Add parts as shown.

STEP 29 Add parts as shown.

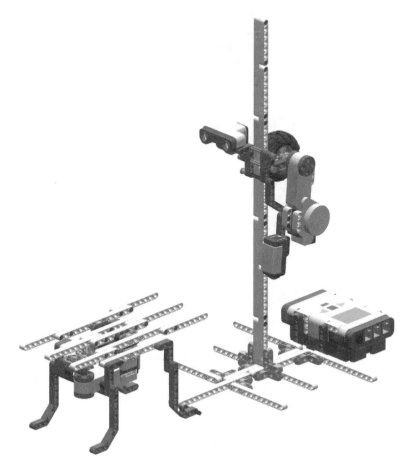

STEP 30 Fix the arm to the base. When you place the scanner head on the support arm, make sure the beams are in the groove so it slides up and down easily.

Now it's time to attach the cables. Connect a long cable between the ultrasonic sensor and port 3. Use another long cable to connect the touch sensor to port 1. Use a medium cable to connect the head motor to port A. Use another medium cable to connect the platform motor to port C.

Programming the Scanner

The tricky part about programming this device is that it needs to know x, y, and z coordinates for every point that it scans. However, the distance sensor only returns one value: the distance to the object. We'll need to use trigonometry to produce coordinates from this value (see Appendix B for trigonometry basics).

The scanner will take a measurement at equal increments all around the object (see Figure 17-3). Notice that our setup produces exactly the same results as moving a sensor around the object. At the second scan (T_2 in Figure 17-3) the effect is the same as if the sensor were moving around the object. It's easier to understand the calculations by imagining the platform standing still and the sensor moving around the object.

We will designate the center of the platform as (0, 0, 0) on our coordinate system. The position of the sensor when it begins scanning will be designated at 0 degrees, and rotation will occur counter-clockwise (the platform rotates clockwise, but from the perspective of the object it appears the sensor moves counter-clockwise).

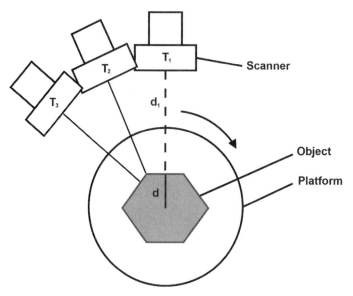

Figure 17-3 Scanning an object

The scanner rotates around the object taking various distance measurements (see Figure 17-4). At T_1 the y coordinate is zero and the x coordinate is the distance from the center of the platform to the position at the edge of the object. The distance from the center of the platform to the sensor (d_t) is constant, and we can easily find it by measuring. We aren't interested in the total distance, however. We want the distance from the center to the edge of the object:

$$d = d_t - d_1$$

At T_2 the x and y coordinates are positive, which can be calculated from d and the angle formed at point A (see Figure 17-4). The equations to calculate all points around the object are as follows:

$$x = d * \sin(A)$$
$$y = d * \cos(A)$$

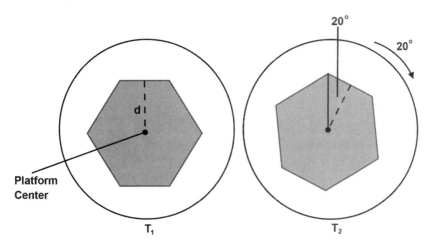

Figure 17-4 Calculating x and y from the rotation angle

The z coordinate is the easiest to calculate, since it is merely the current height of the ultrasonic sensor. Each time the sensor rises, we increase the z coordinate value.

In order for our software to correctly and accurately control and interpret the platform rotation and sensor movement, it needs to know some measurements from the scanner unit: how many degrees the motor must rotate for a single platform rotation, distance from the ultrasonic scanner to the center of the platform, and the distance it raises the scanner after every complete scan. These are given in Table 17-1.

Measurement	Value
Distance from sensor to center	22 cm
Tachometer count for full rotation	12600
Vertical movement per scan line	2.5 cm

Table 17-1 Scanner Unit Constants

Rather than directly measuring the distance from the ultrasonic sensor to the center of the platform, I used the sensor to measure the distance to an object with its edge on the center. The ultrasonic sensor returned 22 centimeters (even though it is closer to 18). It is better to go with the ultrasonic sensor rather than the actual measurement, since all our subsequent values will come from the sensor. With these measurements we can calculate everything else we need to know in order for the scanner to give relatively accurate readings. The full code is as follows:

```
1.  import icommand.nxt.*;
2.
3.  public class ObjectScanner {
4.
5.  // Rotations to Motor.A for one increment:
6.      static final int VERT_INC = 30;
7.  // rotations for full 360 turn of platform:
8.      static final int FULL_TURN = 12600;
9.  // Arbitrary number of samples for each rotation.
10.     static final int TABLE_INC = 20;
11. // degrees to rotate Motor.C for one pixel
12.     static final int ONE_PIXEL = FULL_TURN/TABLE_INC;
13. // Increments to move to top
14.     static final int SCAN_VERT_INC = 5;
15.     static final float CM_PER_VERT_INC = 2.5f;
16. // cm's - dist from sensor to platform center
17. // Use 22 for hard objects, 19 for soft:
18.     static final float DISTANCE_TO_CENTER = 22f;
19.
20.     UltrasonicSensor sonic;
21.     TouchSensor touch;
22.
23.     public ObjectScanner() {
24.         touch = new TouchSensor(SensorPort.S1);
25.         sonic = new UltrasonicSensor(SensorPort.S3);
26.     }
27.
28.     public static void main(String[] args) {
29.         icommand.nxt.comm.NXTCommand.open();
30.
31.         ObjectScanner os = new ObjectScanner();
32.         os.resetScanner();
33.         Point3D [] set = os.fullScan(SCAN_VERT_INC);
34.
35.         icommand.nxt.comm.NXTCommand.close();
36.
37.         for(int i=0;i<set.length;++i) {
38.             System.out.println(i + "\t" + set[i].x + "\t" +
                    set[i].y + "\t" + set[i].z);
39.         }
40.     }
41.
42.     public Point3D[] fullScan(int verticalIncrements) {
43.         Point3D [] pointSet = new Point3D[TABLE_INC *
                verticalIncrements];
44.         for(int i=0;i<verticalIncrements;i++) {
45.             Point3D [] points = scan360(i);
46.             System.arraycopy(points, 0, pointSet,
                    (i * TABLE_INC), points.length);
47.             nextLine();
```

```
48.          }
49.      return pointSet;
50.   }
51.
52.   public Point3D[] scan360(int zLevel) {
53.       Point3D [] pointSet = new Point3D[TABLE_INC];
54.       for(int i = 0;i < TABLE_INC;i++) {
55.          float dist = sonic.getDistance();
56.          dist = DISTANCE_TO_CENTER - dist;// Convert to
                 dist from center of platform
57.
58.          int rotatio = Motor.C.getRotationCount() % FULL_
                 TURN; // !! Don't really need this line because
                    equation handles radians > 2pi, but might
                       as well.
59.          double angle = (double)rotatio / (double)FULL_
                 TURN * 2 * Math.PI;
60.          pointSet[i] = new Point3D();
61.          pointSet[i].x = getX(angle, dist);
62.          pointSet[i].y = getY(angle, dist);
63.          pointSet[i].z = zLevel * CM_PER_VERT_INC;
64.
65.          Motor.C.rotateTo((i+1) * ONE_PIXEL + (zLevel *
                 FULL_TURN));
66.       }
67.      return pointSet;
68.   }
69.
70.   /**
71.    *  Helper method to return x.
72.    * @param angle In radians!
73.    * @param distance In anything (cm or inches)
74.    * @return X coordinate
75.    */
76.   static float getX(double angle, float distance) {
77.       double x = distance * Math.sin(angle);
78.       return (float)x;
79.   }
80.
81.   /**
82.    *  Helper method to return Y.
83.    * @param angle In radians!
84.    * @param distance In anything (cm or inches)
85.    * @return Y coordinate
86.    */
87.   static float getY(double angle, float distance) {
88.       double y = distance * Math.cos(angle);
89.       return (float)y;
90.   }
91.
```

```
92.     public void resetScanner() {
93.         Motor.A.setSpeed(800);
94.         Motor.C.setSpeed(600);
95.         Motor.A.backward();
96.         while(!touch.isPressed()) {}
97.         Motor.A.stop();
98.     }
99.
100.    public void nextLine() {
101.        Motor.A.rotate(VERT_INC);
102.    }
103.
104.    static class Point3D {
105.        float x;
106.        float y;
107.        float z;
108.    }
109. }
```

The VERT_INC constant tells the program how many degrees to rotate the motor each time the sensor moves up. TABLE_INC is the resolution to use for the scan. In other words, it will take 20 samples per rotation. In my tests, 20 samples per rotation produced a respectable result.

Most of the code is self explanatory. The code that actually collects the points is located in scan360(). This method performs one full rotation and produces an array of Points (an inner class).

Using the Scanner

The scanner code outputs a series of 3D coordinates to the console screen. To output these values to a file instead, use the following arguments when running the class:

```
java ObjectScanner >> results.txt
```

Once the program is complete you will see the new file in the same location as your classes. Open the file, then copy and paste the results into one of the applications mentioned below.

The results can be displayed in a variety of ways. Excel is almost ubiquitous to all platforms, so I've included an Excel spreadsheet on the book's website that will display the dots in a 3D image within Excel (see Figure 17-5). If you want something fancier, Windows users can use either DPlot or Graphis to display the images. In DPlot, select File, New 3D Scatter and then select Edit, Paste (Figure 17-6).

WEBSITE: *DPlot trial version:* www.dplot.com
Graphis download: www.kylebank.com

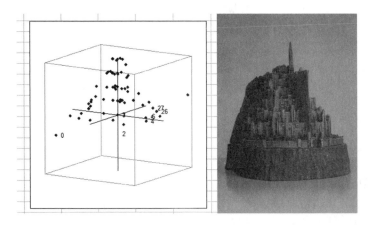

Figure 17-5 Excel displaying the scan of a sculpture

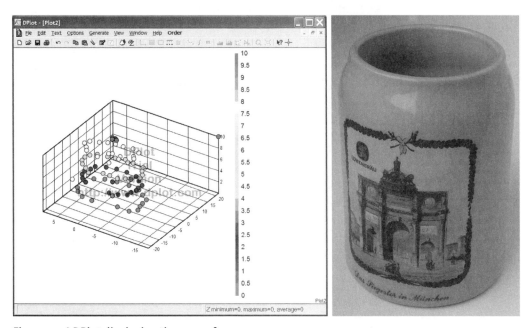

Figure 17-6 DPlot displaying the scan of a mug

The results are actually respectable, considering the limitations of the ultrasonic sensor. The ultrasonic sensor produces measurements in one centimeter increments, which prevents it from picking up fine detail. You can actually change the range of the sensor from 255 centimeters to something smaller and produce more accurate results (see leJOS NXJ API).

The field of view of the ultrasonic sensor is wide, as mentioned previously. If it had a narrower field of view, small details in objects would be detectable. A laser pointer would be ideal.

In practice, the ultrasonic sensor can give off the occasional bad reading. Sometimes it gives readings greater than 100 centimeters, perhaps because the sonar is missing the target. In this case, you will need to edit these values out of your data before displaying them.

TRY IT: Here's a pure programming challenge. Convert the point data into polygons by joining the dots and making triangles. Save it in a format that can be imported into a 3D CAD program for viewing.

Image Scanning

The LEGO RCX did not have a wide variety of commercial sensors, so the light sensor was one of the most popular sensors. As a result, many MINDSTORMS owners built photocopy machines out of LEGO. In this section we will use the NXT light sensor to construct a similar machine.

Building the Photocopier

The goal of the photocopier is to scan an image and display the output to the LCD. It will also try to output the image to a piece of paper. You will need a felt-tipped marker to attach to the scanner.

- Scan the images from a piece of paper
- Display the image on the LCD
- Print the image when the user inserts a piece of paper

STEP 1 Build a long beam from six shorter beams.

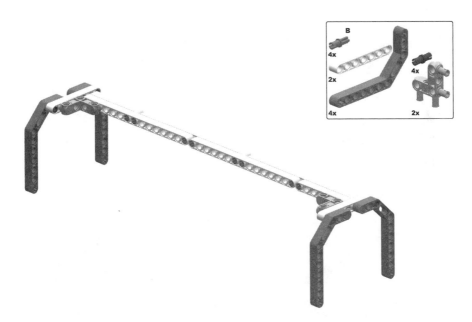

STEP 2 Add parts as shown.

STEP 3 Add rubber connectors to the legs.

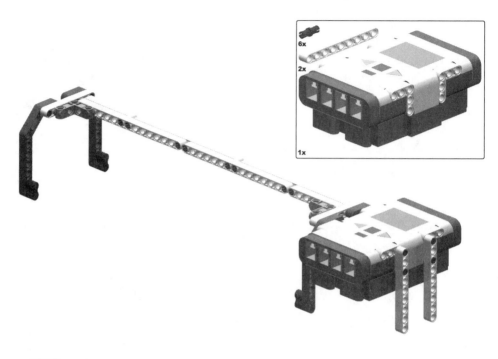

STEP 4 The NXT connects using two pins (obscured).

STEP 5 Add parts as shown.

STEP 6 Add parts as shown.

STEP 7 Add parts as shown.

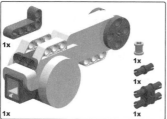

STEP 8 Connect the L-beam using a black pin as shown. Secure the axle with a bush. Connect the motor using the black connector.

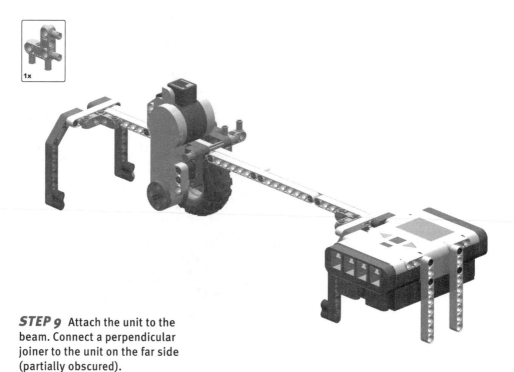

STEP 9 Attach the unit to the beam. Connect a perpendicular joiner to the unit on the far side (partially obscured).

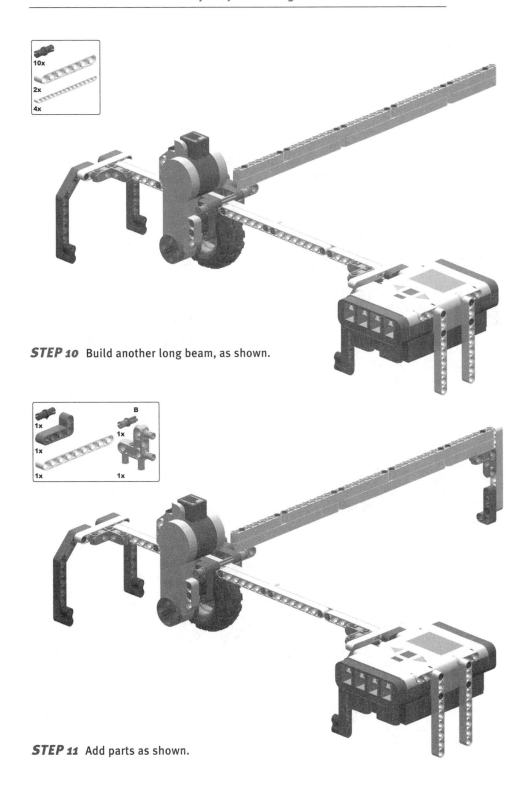

STEP 10 Build another long beam, as shown.

STEP 11 Add parts as shown.

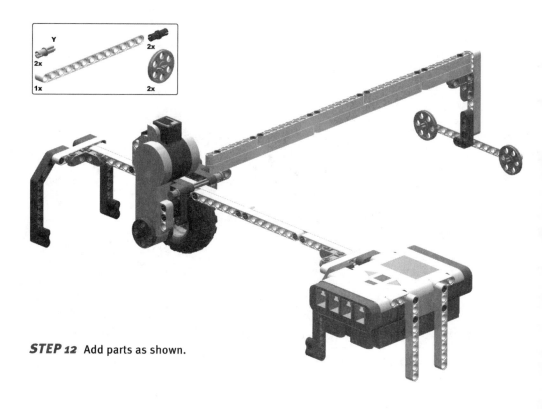

STEP 12 Add parts as shown.

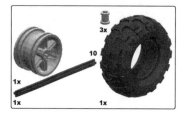

STEP 13 Add parts as shown.

STEP 14 Add parts as shown.

STEP 15 Add parts as shown.

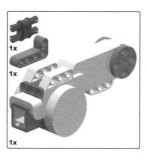

STEP 16 This is a bit disorienting. Rotate the unit 90 degrees clockwise from the previous step. Attach the motor to the wheel axle. Attach the L-beam to the motor using the black connector, as shown.

STEP 17 Rotate the unit back and add the parts as shown.

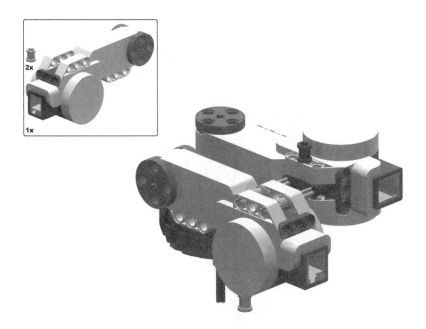

STEP 18 Add parts as shown.

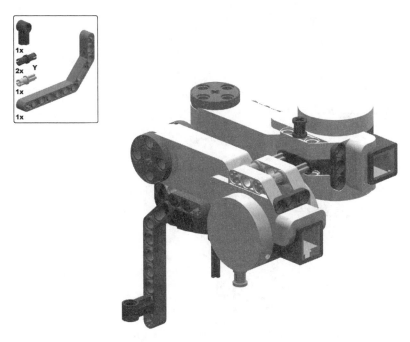

STEP 19 Add parts as shown.

STEP 20 Rotate counter-clockwise. Add the angle connector to the main axle and secure with a bush. Attach the light sensor to the beam with a single pin as shown. Insert the 6-unit axle as shown and secure with two bushes.

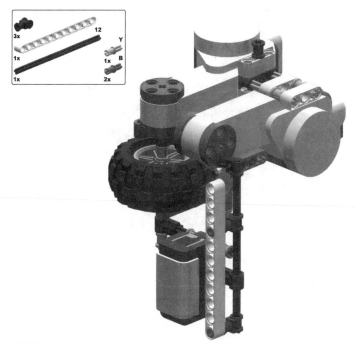

STEP 21 Insert a yellow pin into a motor pin-hole, then add a connector. Add the rest of the parts as shown.

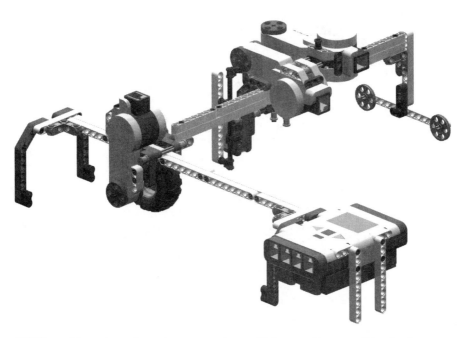

STEP 22 Place the unit on the arm as shown. Make sure the arm slides freely along the beam.

If you want to try outputting the image to a piece of paper, secure a regular felt-tipped marker to the beam next to the light sensor. Using a long cable, connect the light sensor to port 1. Using a medium cable, connect the lone motor to port A. Using another medium cable, connect the pen motor to port B. Connect the final motor to port C using a long cable.

Programming the Photocopier

The photocopier code runs under leJOS NXJ. The code for the photocopier is not complicated since there is very little math involved. It moves the scanner head along the x-axis by one unit, then reads a light value before continuing. Once it gets to the end of the line, it repositions the scanner head and moves the y-axis by one unit. As it scans it draws the image to the LCD screen.

```
1. import lejos.nxt.*;
2. import javax.microedition.lcdui.*;
3.
4. public class Scanner {
5.     int MAX_ROT = 500; // Rotations that covered 21 cm
6.     int MAX_CM = 21; // 21 cm
7.     int ROTATIONS_PER_CM = MAX_ROT/MAX_CM;
8.     int PIXEL_INCREMENT = 10; // Tachocount per pixel
```

```
 9.    LightSensor ls;
10.    int LIGHT_THRESHOLD = 50;
11.    Graphics g;
12.
13.    public Scanner() {
14.        Motor.A.setSpeed(500);
15.        Motor.B.setSpeed(500);
16.        Motor.C.setSpeed(500);
17.        g = new Graphics();
18.        ls = new LightSensor(SensorPort.S1);
19.    }
20.
21.    public static void main(String [] args) {
22.        Scanner s = new Scanner();
23.        int [][] image = s.scanImage(10, 10);
24.        while(!Button.ENTER.isPressed()) {}
25.        s.printImage(image);
26.    }
27.
28.    /**
29.     * @param x_length Length of picture along X axis
            in CM's
30.     * @param y_length Length of picture along Y axis
            in CM's
31.     * @return
32.     */
33.    public int [][] scanImage(int x_length, int y_length) {
34.
35.        int x_count = (x_length * ROTATIONS_PER_CM)/
            PIXEL_INCREMENT;
36.       int y_count = (y_length * ROTATIONS_PER_CM)/
            PIXEL_INCREMENT;
37.        int [][] image = new int[x_count][y_count];
38.        for(int y=0;y<y_count;y++) {
39.            for(int x=0;x<x_count;x++) {
40.                image[x][y] = ls.readValue();
41.                if(image[x][y] < LIGHT_THRESHOLD) {
42.                    g.setPixel(Graphics.BLACK, x, y);
43.                    g.refresh();
44.                }
45.                Motor.A.rotateTo(x * PIXEL_INCREMENT);
46.            }
47.            Motor.C.rotateTo(-y * PIXEL_INCREMENT);
48.        }
49.        return image;
50.    }
51.
52.    public void printImage(int [][] image) {
53.        for(int y=0;y<image[0].length;y++) {
54.            Motor.C.rotateTo(-y * PIXEL_INCREMENT)
```

```
55.          for(int x=0;x<image.length;x++) {
56.              Motor.A.rotateTo(x * PIXEL_INCREMENT);
57.              if(image[x][y] < LIGHT_THRESHOLD)
58.                  Motor.B.rotateTo(180);
59.              else
60.                  Motor.B.rotateTo(0);
61.          }
62.      }
63.   }
64. }
```

The LIGHT_THRESHOLD constant can be adjusted depending on the image you are scanning and your lighting conditions. Currently anything lower than 50% indicates black. If your image is lighter, you might want to raise that threshold value so more pixels are recorded as black.

Using the Photocopier

This robot works best on a smooth table surface. The opposite end of the arm is on wheels which require very little friction; it would lag behind on a carpeted surface.

Adjust the motors to the starting position by grasping each motor near the orange end. Pull it along the double beam until it is close to the NXT brick. Make sure that the beams are fitted into the grooves near each wheel. Place your image on the table and tape it down so it doesn't move, and then start the program.

Once it is finished scanning it will wait for you to press the orange button before it tries to draw the image. Place another piece of paper on top of your original and tape it down securely. Keep in mind the results are not very good, since the drawing mechanism is rather primitive.

TRY IT: HiTechnic produces a color sensor that can be purchased from LEGO. Try modifying this project so it runs under iCommand instead of leJOS NXJ. You can scan a color image and plot it directly to your computer screen, in color. You could also save it as a bitmap file (you will need to research the .BMP file format). Once you have a bitmap, you can even print the result to a color printer.

Behavior-Based Robotics

Topics in this Chapter

- Rodney Brooks' Behavior-Based Robotics
- Subsumption Architecture
- The leJOS NXJ Behavior API

Chapter 18

Behavior-Based Robotics is a branch of robotics programming that uses very little memory and produces insect-level intelligence. The NXT brick has more than enough memory to add interesting behaviors like navigating towards sound, avoiding objects, finding objects, and moving objects. The more behaviors added to your robot, the more interesting it becomes. Let's examine Behavior-Based Robotics.

Behavior-Based Robotics

Behavior-Based Robotics was first defined by Professor Rodney Brooks while at the MIT Artificial Intelligence Laboratory. The strategy of Behavior-Based Robotics is different than most AI programming styles developed before it. The traditional models tend to rely on large data models of the world. They can also be slow to react to changes in the environment. Rodney Brooks took his strategy from the insect world. He noticed that insects are able to perform in the real world with excellent success, despite having very little in the way of memory or intelligence.

If we were to compare an insect to a computer, we would conclude that the insect possessed only a small amount of working memory. After all, it has been shown that insects do not remember things from the past, and can not be trained with Pavlovian methods. So insects rely on a strategy of many simple behaviors that, when alternated with one another, appear as complex behavior. These strategies are effectively hard-wired into the insect and are not learned in the way mammals learn their behavior.

Out of this strategy came the practical implementation of Behavior-Based Robotics, called *subsumption architecture.* This architecture decomposes complex behaviors into several smaller, simple behaviors.

There are essentially two discrete structures that build up a behavior: sensors (inputs) and actuators (outputs). This is true of any organism, not just robots. For example, a mosquito monitors the environment using a variety of sensors, and then reacts. Similarly, if certain sensors are stimulated in a robot, it triggers a reaction. These pairs of conditions and actions are called behaviors (Figure 18-1).

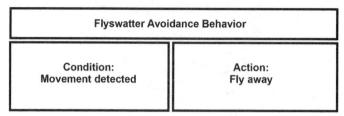

Figure 18–1 Diagram for a Mosquito behavior

Behaviors

Behavior actions can be very simple, such as moving forward, or they can be complex, such as mapping the data in a room. A behavior is just a program the robot (or organism) follows for a period of time. For example, when you want to communicate with someone and are writing an e-mail, you are in letter writing mode, but when your stomach starts to grumble you go to the kitchen and switch over to eating mode. By building up several stimulus/behavior pairs you can theoretically achieve a complex logical model.

Sensors are not restricted to determining when to switch from one mode to another. In fact, they can be used within the action portion of the code as well. If the behavior is to map the boundaries of a room using an ultrasonic sensor, then the action method uses this sensor.

Let's look at a specific example. A firefighting robot could share a single temperature sensor to control more than one reaction. The robot actively searches for areas with greater heat to help it locate fire. If it moves forward and the temperature sensor indicates it is warmer, then it tries another step in that direction. If it gets cooler, then it tries another direction. But assuming there is actually a fire, there will come a time when the temperature gets so high that the robot would melt if it got any closer. To counter this, we could use the same sensor with a higher level behavior to back off if the temperature gets close to the melting point of plastic (Table 18-1).

Behavior	Condition	Action	Priority
Seek heat	Temperature hotter in another area.	Move towards heat.	low
Extinguish flames	Ultraviolet light sensor detects flame	Spray CO_2	medium
Avoid flames	Temperature greater than 160 degrees	Move away from heat	highest

Table 18-1 Firefighting Robot Behavior

A behavior takes over when a condition becomes true. Sometimes a behavior does not have to depend on an external sensor to become true. For example, the robot could monitor an internal condition such as time. When two minutes have elapsed, the condition becomes true. In a way, the timer is a time sensor. Other factors, such as counting the number of bricks a robot has picked up, can also be monitored. Once a robot has collected 10 bricks, a higher level behavior of seeking home and dumping the bricks could become active. Any condition, internal or external to the robot, can activate a behavior.

This doesn't have to be for practical behaviors only. I have seen robots that switch from one emotional mode to another. For example, there is an R2-D2 type of robot that switches from happy mode to bored mode, and sad mode. The bored mode kicks in if the robot has been wandering around for a while without encountering a stimulus. Each of the emotional states is dependent upon what the robot senses in its surroundings, which is really not that different from complex organisms such as humans.

Managing Priorities

There are times when more than one behavior could become activated. In order for a robot to determine which mode gains control, the different behaviors need to have priorities. For example, an animal has several basic goals, such as eating, mating, defending itself, exploring, and protecting its young. Some of these behaviors are more important than others, but all are necessary for the overall survival of the species.

One of the most important behaviors is eating, since starvation prevents an organism from completing any of the other important goals. So we would say eating is a high level priority for an animal. But what if another ferocious animal is attacking it while it is eating? Defending itself (attack mode/retreat mode) would be the highest level behavior for an animal in this situation. Likewise a robot is useless without power, so when the battery level is too low it could seek the recharging station. It is important that the code interrupts current action if a higher level action needs to take over. When this happens, we say that the lower priority behavior has been *suppressed*.

Rodney Brooks developed a standard diagram for representing hierarchies of behaviors (Figure 18-2). A typical diagram shows the behavior, and the order of priorities. When a behavior suppresses other behaviors it is indicated with an S. The rule for Behavior-Based Robotics is that all lower level behaviors are suppressed when a higher level behavior takes over. This means only one behavior can be running at any given time. It might seem that this limits a robot, but organisms really do only one thing at a time. In order to do more than one behavior at a time, a single behavior must include more than one function within it.

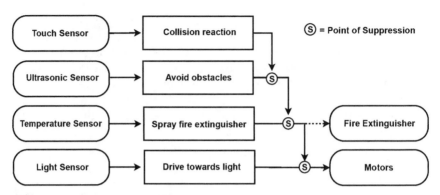

Figure 18–2 Hierarchy of firefighting behavior

The Subsumption API

Many people, when they begin programming robots, think of program flow as a series of if-then statements, which is reminiscent of structured programming. This type of programming is very easy to start but can end up as spaghetti code; all tangled up and difficult to expand.

The Behavior-Based Robotics model, in contrast, requires a little more planning before coding begins, but the payoff is that each behavior is nicely encapsulated within an easily-understood structure. This will theoretically make your code easier for other programmers to understand, and more importantly, make it very easy to add or remove specific behaviors. Let's examine how to do this in leJOS NXJ.

The Behavior API is very simple and is composed of only one interface and one class. The Behavior interface is used to define behaviors. It is very general, since the individual implementations of behaviors vary widely. Once all the Behaviors are defined, they are given to an Arbitrator to regulate. All classes and interfaces for behavior robotics are located in the josx.robotics package. The API for the Behavior interface is as follows.

lejos.subsumption.Behavior

- `boolean takeControl()`

    Returns a boolean value to indicate if this behavior should become active. For example, if a touch sensor indicates the robot has bumped into an object, this method should return true.

- `void action()`

    The code in this method initiates an action when the behavior becomes active. For example, if takeControl() detects that the robot has collided with an object, the action() code could make the robot back up and turn away from the object.

- `void suppress()`

    The code in the suppress() method should immediately terminate the code running in the action() method. The suppress() method can also be used to update any data before this behavior completes.

As you can see, the three methods in the Behavior interface are quite simple. If a robot has two discreet behaviors, then the programmer needs to create two classes, with each class implementing the Behavior interface. Once these classes are complete, the code should hand the Behavior objects to the Arbitrator.

lejos.subsumption.Arbitrator

- `public Arbitrator(Behavior [] behaviors)`

    Creates an Arbitrator object that regulates when each of the behaviors will become active. The higher the index array number for a Behavior, the higher the priority level.

    Parameters
    `behaviors:` An array of Behaviors.

- `public void start()`

    Starts the arbitration system.

The Arbitrator class is even easier to understand than Behavior. When an Arbitrator object is instantiated, it is given an array of Behavior objects. Once it has these, the start() method is called and it begins arbitrating – deciding which behaviors should become active. The Arbitrator calls the takeControl() method on each Behavior object, starting with the object with the highest index number in the array. It works its way through each of the behavior objects until it encounters a behavior that wants to take control. When it encounters one, it executes the action() method of that behavior once and only once. If two behaviors both want to take control, then only the higher level behavior will be allowed (Figure 18-3).

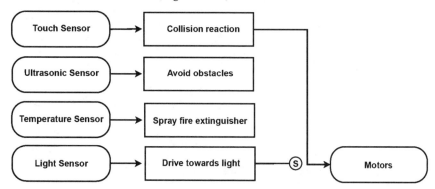

Figure 18–3 Higher level behaviors suppress lower level behaviors

Programming Behavior-Based Robotics

Now that you are familiar with the Behavior API under leJOS, let's look at a simple example using three behaviors. For this example, we will program some behavior for R2MeToo (Chapter Two). We'll give R2MeToo three simple behaviors:

1. Drive forward

2. If too close to an object, find a new direction.

3. If wheels seize up (undetected collision) back out.

The behavior hierarchy is displayed in Figure 18-4. Notice that instead of controlling motors directly, the behaviors access a shared Pilot object. This makes it easier to control movement, rather than going straight to the motors.

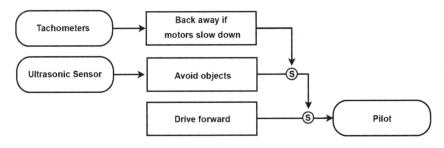

⑤ = Point of Suppression

Figure 18-4 R2MeToo behavior hierarchy

Let's start with the first behavior. As we saw in the Behavior interface, we must implement the methods action(), suppress(), and takeControl(). The behavior for driving forward will take place in the action() method. It simply needs to make the robot drive forward. Next, the suppress() method will need to stop this action when it is called.

```
1.  import lejos.subsumption.*;
2.  import lejos.nxt.*;
3.  import lejos.navigation.*;
4.
5.  public class BehaviorForward implements Behavior {
6.
7.      Pilot robot;
8.
9.      public BehaviorForward(Pilot p) {
10.         this.robot = p;
11.     }
12.
13.     public boolean takeControl() {
14.         return true;
15.     }
```

```
16.
17.    public void suppress() {
18.        robot.stop();
19.    }
20.
21.    public void action() {
22.        robot.forward();
23.    }
24. }
```

That's all it takes to define our first Behavior to drive the robot forward. This robot will always drive forward, unless something suppresses the action, so this Behavior should always want to take control (it's a bit of a control freak). The takeControl() method should return true, no matter what is happening. This may seem counter intuitive, but rest assured that higher level behaviors will be able to cut in on this behavior when the need arises.

The second behavior is a little more complicated than the first, but similar. The main action of this behavior is to reverse and turn when the robot comes within 40 centimeters of an object. The complete listing for this behavior is as follows:

```
1. import lejos.nxt.*;
2. import lejos.subsumption.*;
3. import lejos.navigation.*;
4.
5. public class BehaviorProximity implements Behavior {
6.     UltrasonicSensor us;
7.     Pilot robot;
8.
9.     public BehaviorProximity(UltrasonicSensor us, Pilot p) {
10.        this.us = us;
11.        this.robot = p;
12.    }
13.
14.    public boolean takeControl() {
15.        int dist = us.getDistance();
16.        return (dist < 40);
17.    }
18.
19.    public void suppress() {
20.        robot.stop();
21.    }
22.
23.    public void action() {
24.        robot.stop();
25.        int bestDir = getBestDir();
26.        robot.rotateTo(bestDir);
27.    }
28.    // Rotate head and find longest direction
```

```
29.     public static int getBestDir() {
30.         ls.setFloodlight(true);
31.         int bestDir = 0;
32.         int bestDist = 0;
33.         for(int i=-SWEEP/2;i<SWEEP/2;i = i + INTERVAL) {
34.             Motor.A.rotateTo(i * 9); // 9 = gear ratio
35.             int curDist = us.getDistance();
36.             if(curDist > bestDist) bestDir = i;
37.             if(curDist >  200) break;
38.         }
39.         Motor.A.rotateTo(0);
40.         ls.setFloodlight(false);
41.         return bestDir;
42.     }
43. }
```

Now that we have two of the behaviors defined, we need to create the main method using an Arbitrator:

```
1. import lejos.robotics.*;
2. import lejos.nxt.*;
3.
4. public class BehaviorMain {
5.     static final float WHEEL_DIAM = 5.6F;
6.     static final float TRACK_W = 13F;
7.
8.     public static void main(String [] args) {
9.         UltrasonicSensor us = new UltrasonicSensor
                (SensorPort.S1);
10.        Pilot robot = new Pilot(WHEEL_DIAM, TRACK_W,
                Motor.C, Motor.B, true);
11.        robot.setSpeed(500);
12.        Behavior b1 = new BehaviorForward(robot);
13.        Behavior b2 = new BehaviorProximity(us, robot);
14.        Behavior [] bArray = {b1, b2};
15.        Arbitrator arby = new Arbitrator(bArray);
16.        arby.start();
17.     }
18. }
```

The code above is easy to understand. The first two lines in the main() method create the ultrasonic sensor and pilot objects. The next lines instantiate the Behaviors. Note that the Behaviors share the Pilot object. This is perfectly legal in subsumption architecture. The next line places the Behaviors into an array, with the lowest priority Behavior taking the lowest array index. The next line creates the Arbitrator, and the final line starts the Arbitration process.

This is a lot of work for two simple behaviors, but now let's see how easy it is to insert a third behavior without altering any code in the other classes. This is the part that makes subsumption architecture very appealing for robotics programming.

The third behavior is more complicated because it will monitor the motor rotation and attempt to detect a slowdown in speed:

```
1.  import lejos.subsumption.*;
2.  import lejos.nxt.*;
3.  import lejos.navigation.*;
4.
5.  public class BehaviorCollision implements Behavior {
6.
7.      boolean isUpToSpeed = false;
8.      Pilot robot;
9.
10.     public BehaviorCollision(Pilot p) {
11.         robot = p;
12.     }
13.
14.     public boolean takeControl() {
15.
16.         if(Motor.B.isStopped()) isUpToSpeed = false;
17.         if(Motor.B.isRotating()) isUpToSpeed = false;
18.         if(Motor.B.getSpeed() < 0) isUpToSpeed = false;
19.
20.         int targetSpeed = Motor.B.getSpeed();
21.         int realSpeed = Motor.B.getActualSpeed();
22.
23.         if(realSpeed >= targetSpeed) isUpToSpeed = true;
24.         if(isUpToSpeed)
25.             return realSpeed < ((targetSpeed * 95) / 100);
26.         else
27.     return false;
28.     }
29.
30.     public void suppress() {
31.         robot.stop();
32.     }
33.
34.     public void action() {
35.         robot.travel(-75);
36.         robot.rotate(-80);
37.     }
38. }
```

The action() and suppress() methods above are very simple, but the takeControl() method requires some explanation. The code is very careful not to falsely detect a slowdown in speed. It uses a variable flag to make sure it allows the motor to read the target speed before it begins searching for a slowdown, otherwise it would think the motors are seizing as the robot starts moving and takes some time to reach full speed. Anytime the motor is stopped, is reversed, or is executing a tachometer rotation, the flag is reset.

Inserting the Behavior into our scheme is a small task. We simply alter the code of our main class as follows:

```
Behavior b3 = new BehaviorCollision(robot);
Behavior [] bArray = {b1, b2, b3};
```

This beautifully demonstrates the real benefit of Behavior-Based Robotics coding. Inserting a new behavior is simple; it is grounded in object-oriented design. Each behavior is a self contained, independent object.

TIP: When creating a Behavior-Based Robotics system, it is best to program the behaviors one at a time and test them individually. If you code all the behaviors and then upload them all at once to the NXT brick, there is a good chance a difficult to locate bug will exist somewhere in the behaviors. By programming and testing them one at a time it will be easier to identify where a problem was introduced.

Results

Subsumption architecture is an interesting programming architecture that has potential, but also some recognized problems. The main disadvantage of this model is that the more layers you add to a program, the more likely it becomes that the behaviors will interfere with one another.

The promise is that you will get interesting behavior from your robot, but you are limited by two physical considerations: sensors and motors. Most robots just roll around, so how much interesting behavior can be produced? You can go forward, backward, or turn.

The sensors need to be varied too. A touch sensor can do one thing – indicate pressed or not. A light sensor is better because it has variation, but the reactions to it are limited. The more sensors you have, the more varied reactions can become.

It would be nice if all behaviors were as simple as the examples given above, but in more complex coding unexpected results can sometimes be introduced. Threads, for example, can sometimes be difficult to halt from the suppress() method. This can lead to two different threads fighting over the same resources – often the same motor! Another potential problem that can occur in multi-threaded programs is events such as touch sensor hits going undetected.

Since behaviors are totally suppressed when a higher level takes over, robots tend to do one thing at a time rather than multitasking. For example, if both motors are moving forward and a higher level behavior takes command it is not clear if all lower level motor movements should be stopped. What if the higher level behavior only uses one of the motors? Should the other keep moving forward? And will this lead to odd behavior?

So why use the Behavior API? The best reason is because in programming we strive to create the simplest, most powerful solution possible, even if it takes slightly more time. The importance of reusable, maintainable code has been demonstrated repeatedly in the workplace, especially on projects involving more than one person. If you leave your code and come back to it several months later, things that looked clear suddenly are no longer so obvious. Behavior-Based Robotics allows you to add and remove behaviors without even looking at the rest of the code.

Another big plus of Behavior-Based Robotics is that programmers can easily exchange Behaviors with each other, fostering code reusability. Dozens of interesting, generic behaviors could be uploaded to websites, and you could simply pick the behaviors you want to add to your robot, assuming the behaviors use generalized objects like Pilot and Navigator.

Expanding the NXT

Topics in this Chapter

- Using RCX sensors with NXT
- Connecting up to four RCX motors to the NXT
- Creating an RCX slave
- Expanding NXT Ports

Chapter 19

Sometimes you need an extra motor or sensor to get the job done. Like many NXT owners, you might already own a Robotics Invention System, which contains two motors, two touch sensors and one light sensor. If only you could use those parts in your NXT robots! This chapter will show you how to interface RCX parts with the NXT brick.

Fortunately, third party developers have created adapters to allow you to use additional motors and sensors. These include:

- Cables that adapt an RCX connector to an NXT port.
- A device that converts a single NXT port into four RCX connectors.
- A wireless device that allows the NXT to control the RCX brick as a slave.
- Devices that add more sensor and motor ports to the NXT

Using RCX sensors with NXT

The NXT kit contains one of every sensor, but sometimes you may need two or more of the same sensor for a project. The RIS kit contains two touch sensors and a light sensor, plus there are temperature sensors and rotation sensors available for the RCX. If you need two of the same type of sensor, it can be very helpful to connect an RCX sensor to your robot.

LEGO markets a cable with a 2 x 2 RCX brick connector on one end and an RJ12 plug on the other (see Figure 19-1). This cable allows you to use all the RCX sensors and motors with the NXT. You can even use some of your NXT sensors on the RCX. To order this cable, please refer to Appendix A.

Figure 19-1 The RCX to NXT cable

For the most part, the legacy RCX sensors use the same classes as the NXT sensors, so you don't have to do anything special in your leJOS code to access them. Only the light sensor differs, so you must use the RCXLightSensor class when using the old sensor.

Changing a Port into 4 RCX Connectors

Mindsensors.com offers a motor multiplexer that transforms a sensor port into four RCX motor ports (see Figure 19-2).

Figure 19-2 RCX Motor Multiplexer (© 2007 Mindstorms.com)

Unfortunately, you also need an external power source for the multiplexer. A power supply that supplies 9 volts is ideal, though you can use any DC supply up to 35 volts. A 9 volt battery works but it does not last as long as multiple AA batteries. If the robot is stationary, a DC transformer will work (5 volts is acceptable, 9 volts is better). The LEGO battery box (see Appendix A) is an ideal option for long run times and portability. The LEGO battery box plugs directly into the port (see arrow in Figure 19-2). If you want to connect a different supply like a 9V battery or transformer, you must connect two wires to the battery's positive and negative terminals, and then plug the wires into the appropriate terminals on the multiplexer.

There are downsides to using RCX motors, of course. The RCX motors have no tachometer, so this might limit their usefulness. You will probably end up using them for control of secondary devices on your main robot platform, such as rotating the ultrasonic sensor. If you need precise movements, they will have to be calibrated using either a touch sensor, a rotation sensor, or possibly a light sensor.

Battery constraints limit each motor to running only at about medium power, which means you will have to convert the movement into slow but stronger movement by gearing down. This can add to the size and complexity of a robot.

This adapter works out well for many robot projects. You can use the three NXT ports for NXT motors and sacrifice only a single sensor port to gain up to four additional RCX motors. This gives you a total of seven motors and three sensors, enough to add a robot arm and hand to a mobile robot!

The leJOS NXJ API for using the multiplexer is located in the class lejos.nxt.RCXMotorMultiplexer. It's similar to using the Motor class, with methods to set speed, get speed, and control the directions of motors. The following code demonstrates how to use RCXMotorMultiplexer:

1. `RCXMotorMultiplexer multi = new RCXMotorMultiplexer (SensorPort.S1);`
2. `multi.A.setSpeed(255);`
3. `multi.A.forward();`

NOTE: *To view all the methods in the RCXMotorMultiplexer class, view the API documentation included with leJOS or visit* **www.lejos.org** *and* click the API link on the left side.

Making an RCX Slave

Lego was in a bit of a bind when designing the NXT. If they included a fourth motor port, where does the USB port fit? As you can see there is really no room left. If they included four motor ports and three sensor ports, they would be expected to include four motors, otherwise a few people would invariably complain that they were forcing people to buy a fourth motor.

The cheapest way for RCX owners to gain more motors is to use their RCX brick for additional ports. This is accomplished by using a commercial NXT to RCX adapter. There are two available, the NRLink from Mindsensors.com (see Figure 19-3A) and the iRLink by HiTechnic (see Figure 19-3B).

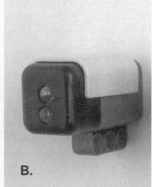

Figure 19-3 IR NXT to RCX adapters

The downside of this solution is that your robot must accommodate the RCX brick with six AA-batteries. And of course, like the previous solution, the RCX motors are not servos. Furthermore, RCX parts use

438

traditional studded bricks for construction, making them difficult to integrate into your NXT robot.

That aside, there are many advantages. You get three more motor ports and three more sensor ports at full power. Control of these new devices integrates nicely in leJOS, making them easy to program. Let's look at some sample code that runs from either on the NXT brick or from iCommand:

```
1. RCXLink rcx = new RCXLink(SensorPort.S1);
2. rcx.A.forward();
3. rcx.A.stop();
4. rcx.beep();
5. rcx.runProgram(0); // Run program 0 from RCX
6. rcx.powerOff();
```

For this code to work, the IR adapter must be connected to NXT port 1 and pointed at the RCX IR port (see Figure 19-4). The RCX brick needs to have the standard LEGO firmware uploaded, not a replacement firmware like leJOS.

Figure 19-4 RCX to NXT IR linking

Not only can the RCX act as a slave, but it can also be a totally autonomous robot that merely exchanges data with the NXT. For example, an NXT robot could act like a Saint Bernard dog by looking for a lost RCX robot in another room. When it finds the RCX, it could tell it how to find its way home by sending some helpful coordinates.

NOTE: *To view all the methods in the RCXLink class, view the included API documentation that comes with leJOS or visit* www.lejos.org *and click the API link on the left side.*

Expanding NXT Ports

One of the most valued expansions for the NXT brick is to gain extra motor ports. Imagine having seven fully powered NXT servo motors instead of only three. You could create vastly more interesting designs such as complex walking robots or mobile robots that can manipulate objects.

Both HiTechnic and Mindsensors.com offer motor multiplexers for the NXT. HiTechnic calls theirs the mMux (Motor Multiplexor) and Mindsensors.com calls theirs the NXT Motor Extender (see Figure 19-5).

Figure 19-5 NXT Motor Extender (Mindsensors.com)

These multiplexers connect to an input port on the NXT brick, so one sensor port is sacrificed for four motor ports. They also require an external power supply such as the LEGO battery box (see Appendix A).

HiTechnic also has a sensor multiplexer, which they appropriately call the sMux. All the power to the sensors is supplied from the NXT brick since sensors require less power than motors. The multiplexer works with I²C sensors (ultrasonic and compass) and regular analog sensors (light sensors).

These multiplexers are in the prototype stage as of this writing. When they are released, the leJOS developers will create classes and methods for them, making it easy to control the extra motors as though they were regular motors.

WEBSITE: *For further information on the multiplexers discussed above, visit the respective company websites:* www.hitechnic.com *and* www.mindsensors.com

GPS & Harsh Terrain

Topics in this Chapter

Chapter 20

Although the NXT is designed for indoor use, you can create LEGO robots that are suitable for the outdoors. The first obstacle to overcome is rough, uneven terrain. Once you overcome this, you can try keeping track of coordinates using a GPS receiver, which works well outdoors.

GPS Localization

GPS is a tempting solution to the localization problem for a few reasons. First, it is widely available and relatively easy to use with the LEGO NXT. Second, it has no drift problems. As we saw in earlier chapters, whether using tachometers or a compass for navigation, after a while the robot was no longer accurately reporting its position. GPS takes constant new readings, so there is no accumulation of errors.

Unfortunately, as GPS exists today it is not good for small mobile robots. First, it doesn't work very well indoors. The GPS receiver needs to receive a signal from at least three satellites to triangulate a position. Second, GPS has a margin of error of about two to five meters for most GPS units. This is fine for large vehicles and humans operating in wide open spaces over long distances. However, when you are only four inches tall and navigating in a few square meters, it doesn't prove very helpful. Third, each GPS reading takes several seconds to triangulate. This means you don't get near-continuous updates of the robot position. It won't work very well if your robot is on the move.

NOTE: *I don't recommend purchasing a GPS system just to use with NXT. However, if you already own one it might be worth experimenting with.*

This section will demonstrate GPS and how to access coordinates through program code. Even though it isn't helpful for navigating, you can use it for other NXT applications, such as driving around a racetrack collecting accelerometer data. Or, if your robot is able to handle rough terrain, you can try an outdoor robot.

Programming GPS

For this project you will need a GPS device for a PDA using Windows Mobile. Optionally, you can connect the GPS to your computer through a serial port. A Bluetooth GPS device, such as the one shown in Figure 20-1, is handier if you are using a computer since it is more portable.

Figure 20-1 Trimble Bluetooth GPS

NOTE: As of this printing, most GPS software is written to communicate through serial ports. If your GPS device uses USB, you can try using a USB to serial adapter (provided your computer has a serial port). Bluetooth works because is uses serial port profile (SPP).

Programs can interact with GPS devices using a variety of protocols, but the standard protocol supported by almost all GPS devices is NMEA. This protocol was developed by the National Marine Electronics Association for use with marine devices, such as navigation instruments, depth finders, and of course GPS devices.

When a GPS is set to NMEA mode, it continuously outputs data from the GPS unit. A typical sentence is shown below. Most people are interested in obtaining the latitude and longitude from the GPS unit, so we will examine one of these sentences:

`$GPGLL,4951.3637,N,09706.1689,W,185113.203,A*2E`

All NMEA messages from the GPS device begin with $GP (sentences starting with $P indicate proprietary sentences by the manufacturer). The next three letters, GLL, indicate that the sentence is a GPS-derived latitude and longitude. Latitude and longitude are expressed as degrees, minutes, and seconds (these are not time values as on a clock, but they range from 0 - 60). Latitude can be anywhere up to 90 degrees, while longitude is up to 180 degrees. So the above reading translates to the following:

Latitude: 49° 51 36.37 N
Longitude: 97° 06 16.89 W

The next number is the Coordinated Universal Time stamp (UTC), and the final number indicates whether the data is valid (A means it is valid).

TRY IT: *You can view NMEA messages live from your GPS by using a terminal program. In this example we will use Hyperterminal.*

1. Connect your GPS unit to a serial port.
2. Select Start > All Programs > Accessories > Communications > Hyperterminal.
3. The window will ask for a Connection Description. Type in GPS Test.
4. Another dialog will appear. Where it says *Connect using* select your com port (see Figure 20-2). Normally this is COM1 if you are plugging into the com port directly, or COM4 to COM7 if you are using a Bluetooth GPS.

Figure 20-2 Selecting the COM port

5. It will then ask for the com port properties. Set Bits per second to 4800, 8 data bits, no parity, 1 stop bit, and no flow control (see Figure 20-3). Click OK.

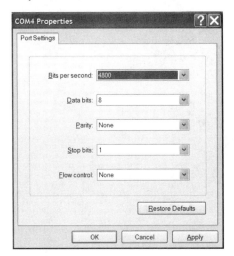

Figure 20-3 Choosing the COM port settings

6. You will immediately see a constant flow of NMEA sentences from the GPS (see Figure 20-4).

```
GPS Test - HyperTerminal
File  Edit  View  Call  Transfer  Help

$GPAPB,A,A,0.1,R,N,,,213.7,T,SIM003,216.1,T,,,*3C
$GPGSA,A,3,01,02,03,04,,,,,,,,2.0,2.0,2.0*34
$GPGSV,3,1,12,17,80,324,,24,69,104,,21,68,294,,26,56,108,*73
$GPGSV,3,2,12,06,39,192,,13,37,164,,18,37,243,,04,23,107,*73
$GPGSV,3,3,12,27,12,070,,14,08,193,,08,02,012,,03,01,327,*70
$GPGLL,4951.4687,N,09706.1141,W,185050.812,A*2D
$GPGGA,185050.81,4951.4687,N,09706.1141,W,1,04,2.0,00026,M,,,,*38
$GPRMB,A,0.12,R,SIM002,SIM003,4949.2246,N,09708.6538,W,002.8,216.,021.7,V*0F
$GPRMC,185050.81,A,4951.4687,N,09706.1141,W,21.7,219.1,260307,06.,E*7A
$GPAPB,A,A,0.1,R,N,,,213.7,T,SIM003,216.1,T,,,*3C
$GPGSA,A,3,01,02,03,04,,,,,,,,2.0,2.0,2.0*34
$GPGSV,3,1,12,17,80,324,,24,69,104,,21,68,294,,26,56,108,*73
$GPGSV,3,2,12,06,39,192,,13,37,164,,18,37,243,,04,23,107,*73
$GPGSV,3,3,12,27,12,070,,14,08,193,,08,02,012,,03,01,327,*70
$GPGLL,4951.4621,N,09706.1175,W,185052.219,A*25
$GPGGA,185052.22,4951.4621,N,09706.1175,W,1,04,2.0,00026,M,,,,*38
$GPRMB,A,0.12,R,SIM002,SIM003,4949.2246,N,09708.6538,W,002.8,216.,021.7,V*0F
$GPRMC,185052.22,A,4951.4621,N,09706.1175,W,21.7,219.1,260307,06.,E*7A
$GPAPB,A,A,0.1,R,N,,,213.7,T,SIM003,216.2,T,,,*3F
$GPGSA,A,3,01,02,03,04,,,,,,,,2.0,2.0,2.0*34
$GPGSV,3,1,12,17,80,324,,24,69,104,,21,68,294,,26,56,108,*73
$GPGSV,3,2,12,06,39,192,,13,37,164,,18,37,243,,04,23,107,*73
$GPGSV,3,3,12,27,12,070,,14,08,193,,08,02,012,,03,01,327,*70

Connected 0:00:10    Auto detect    4800 8-N-1    SCROLL    CAPS    NUM    Capture    Print echo
```

Figure 20-4 The GPS outputting NMEA sentences

If you receive garbled text, your baud rate is probably not set to 4800, either on the GPS or on your terminal. If nothing appears, you've probably selected the wrong COM port.

You can parse these values directly from the GPS unit with Java. First you need an input stream from the serial port. Java has two popular APIs for receiving data from serial ports. First is the official Java Communications API by SUN, however they are no longer updating this package for Windows. The other option is RXTX, an open source API for multiple platforms. Both of these packages use the same class names and methods, making them easily interchangeable.

WEBSITE:
Sun Java Communications API:
java.sun.com/products/javacomm/
RXTX: www.rxtx.org

The MySaifu JVM for portable devices has its own implementation of the Java communications API, so if you are using a Pocket PC to interact with the GPS, this is the solution you should use.

Automatic NMEA Parsing

If you don't want to manually parse NMEA data from your GPS, you can use classes from other packages. A program called GPSylon contains classes for acquiring data from NMEA streams.

Installation:

1. Visit gpsmap.sourceforge.net and download the GPSylon package. You will probably want the binary distribution, unless you want to study the source code.

2. Extract into a directory. e.g. c:\java\gpsylon

3. Add gpsylon.jar to your classpath. The jar file is in the main GPSylon directory. If you are using Eclipse, make sure to add this package to Project > Properties > Java Build Path. Don't forget to add the native library location of RXTX under this entry in Eclipse.

   `c:\java\gpsylon\gpsylon-0.5.2.jar`

4. You also need to add RXTX to your classpath. It's located in \lib\java. There are also some native libraries that RXTX relies on. Make the dlls accessible by copying them to your Java bin directory e.g. c:\j2sdk1.4.2_12\bin.

Now that Java has access to the GPSylon classes, the code to read GPS information is fairly straightforward:

```
1.  import org.dinopolis.gpstool.gpsinput.*;
2.  import org.dinopolis.gpstool.gpsinput.nmea.*;
3.  import java.util.Hashtable;
4.  import java.beans.*;
5.
6.  public class GPSTest {
7.
8.      public static void main(String [] args) {
9.          GPSDataProcessor nmeaReader = new
                  GPSNmeaDataProcessor();
10.         Hashtable settings = new Hashtable();
11.         // Linux users use something like: /dev/ttyS1
12.         settings.put(GPSSerialDevice.PORT_NAME_KEY, "COM1");
13.         settings.put(GPSSerialDevice.PORT_SPEED_KEY, new
                Integer(4800));
14.         GPSDevice myGPS = new GPSSerialDevice();
15.         try {
16.             myGPS.init(settings);
17.             nmeaReader.setGPSDevice(myGPS);
18.             nmeaReader.open();
19.         } catch (GPSException e) {
20.             System.out.println("Error: " + e);
21.         }
22.
23.         PropertyChangeListener gps_listener = new
                PropertyChangeListener() {
```

```
24.        public void propertyChange (PropertyChange
              Event e) {
25.            Object obj = e.getNewValue();
26.            String propertyType = e.getPropertyName();
27.            if(propertyType.equals(GPSDataProcessor.
              LOCATION)) {
28.              GPSPosition pos = (GPSPosition)obj;
29.              System.out.print("Latitude: " + pos.
                getLatitude());
30.              System.out.println("  Longitude: " + pos.
                getLongitude());
31.            }
32.          }
33.        };
34.      nmeaReader.addGPSDataChangeListener(GPSData
            Processor.LOCATION, gps_listener);
35.    }
36. }
```

Windows users should make sure to change the serial port to the proper value, such as COM1. Windows Mobile users must append colons (COM6:) to the end of the serial port. Linux and Mac users must change the serial port to the form /dev/ttyS1.

If you are trying to use the above code on a Pocket PC, you will need to change the import statement in one class to use the Java Communications API instead of RXTX. Download the source file from the GPSylon site. The class that requires changes is org.dinopolis. gpstool.gpsinput.GPSSerialDevice.

Replace the following:
```
1. import gnu.io.CommPortIdentifier;
2. import gnu.io.SerialPort;
3. import gnu.io.PortInUseException;
4. import gnu.io.UnsupportedCommOperationException;
5. import gnu.io.NoSuchPortException;
```

With:
```
1. import javax.comm.CommPortIdentifier;
2. import javax.comm.SerialPort;
3. import javax.comm.PortInUseException;
4. import javax.comm.UnsupportedCommOperationException;
5. import javax.comm.NoSuchPortException;
```

Compile this class and replace the old class in gpsylon.jar with your new class. You can use a compression tool such as Winzip or Winrar to open the jar file and replace the class. Upload your new jar file to your Pocket PC and add it to the classpath in Mysaifu.

Alternate Java GPS Solutions

SuperWaba is a JVM for portable devices. I find it more complicated to use than MySaifu because the class files must be converted to a proprietary file type to run on portable devices (MySaifu runs class files directly). However, the SuperWaba API includes GPS classes:

www.superwaba.com.br

There is also a full featured Java GPS API from Chaeron. It uses parts of SuperWaba in order to use GPS:

www.chaeron.com/gps.html

Overcoming Harsh Terrain

Normally you would think a wheeled robot with two inch diameter wheels would be incapable of climbing over a vertical obstacle like a step. Intuition tells you it would just bump up against it and stop. However, there's a special type of suspension called Rocker-Bogie that can surmount vertical objects larger than the wheels.

The name Rocker-Bogie was coined by inventor Don Bickler. Rocker comes from the wheel suspension, which uses several rockers (see-saws) to cause one set of wheels to lower when the other set is raised. Bogie mimics the device used in trains to keep all the wheels on the track at the same time. A train is actually capable of going over rough track without derailing because of the sophisticated bogie suspension. While working at the Jet Propulsion Laboratory, Bickler adapted the four-wheeled bogie system into a 6-wheeled Rocker-Bogie system and placed a patent on it in 1989 (see Figure 20-5).

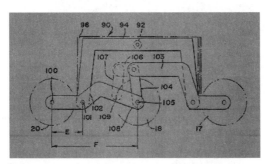

Figure 20-5 The original Rocker-Bogie patent

Rocker-Bogie suspension was used in the first tiny Mars rover, Sojourner (Figure 20-6), and later with the 185 kg MER, Mars Exploration Rover. The Rocker-Bogie system employed by Sojourner and MER are simplified versions of the design patented by Bickler.

Figure 20-6 Sojourner exploring Mars

Building the Rocker-Bogie Suspension

Rocker-Bogie suspension must be seen to be believed, so we'll try building it with LEGO. Going into this project I knew it would be difficult to build a 6-wheel drive Rocker-Bogie system for a number of reasons. First, the NASA Sojourner uses tiny motors on the front and rear wheels to steer – four motors in total, just for steering. The Sojourner also uses an additional six motors to drive the vehicle – one for each wheel. That's ten motors in total. Our goal is to build this robot with only three motors.

There's also a very obvious limitation on wheels. The NXT kit contains four wheels and we need six. This can be overcome by using large gears for two of the rear wheels, but if you have an RIS kit, there are some excellent tires to use as replacements.

Finally, there were a limited number of long axles and bevel gears to transfer the axis of rotation along the supports of the robot. In the end, there were only enough to support the front four wheels. There are two options for the rear wheels: either join the two rear wheels with a single axle (which reduces the functionality of the suspension system) or make it five-wheel drive instead of six.

The goals for the robot are as follows:

- Powerful drive system by gearing down the servos even more.
- Six wheel drive.
- True Rocker-Bogie system to handle rough terrain.
- Enough clearance to climb steps.

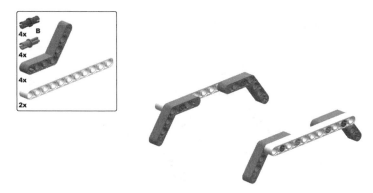

STEP 1 Add parts as shown.

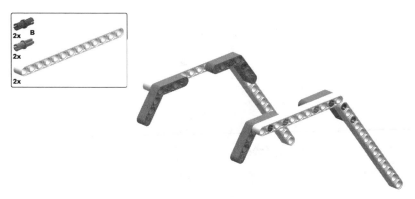

STEP 2 Add parts as shown.

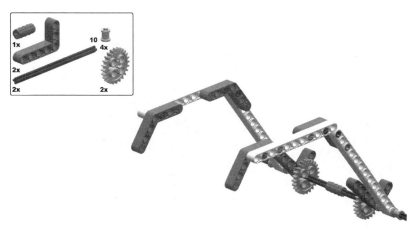

STEP 3 Place all parts on the axles as shown. The L-beams hang loosely for now. The bushes go on either side of the white beams. Leave about one and a half units on the end of the axles.

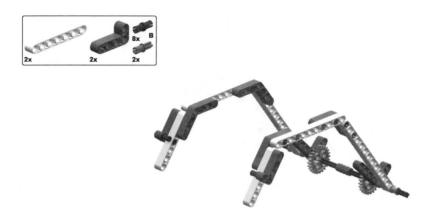

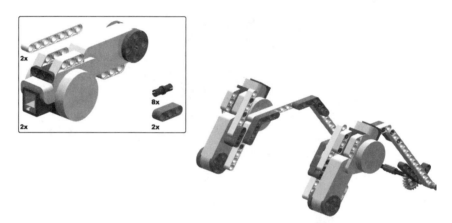

STEP 4 Attach the white beams using one black pin and one blue axle-pin. Attach the L-beams using two black pins as shown. Add the remaining pins.

STEP 5 Attach motors as shown. They hang loosely for now.

STEP 6 Add parts as shown.

STEP 7 Attach both angle connectors using 3-unit pins. Insert the 7-unit axle through the angle connectors and secure with bushes.

STEP 8 Add parts as shown.

STEP 9 Attach the 3-unit beams to the triangles. Loosely connect them to the center axle.

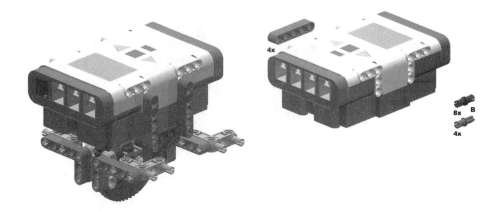

STEP 10 After this step, the brick will see-saw freely.

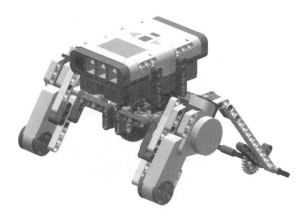

STEP 11 Attach the brick to the main unit as shown.

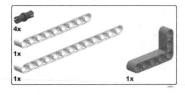

STEP 12 Add parts as shown.

STEP 13 Add parts as shown.

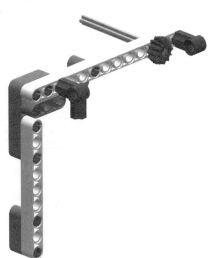

STEP 14 Add parts as shown.

STEP 15 Add parts as shown.

STEP 16 Attach all parts to the axle as shown. The bush is attached on the other side to hold the axle in place (obscured).

STEP 17 Add parts as shown.

STEP 18 The axle that attaches to the knob-wheel is 8 units, but we really need a 9-unit axle. The 8-unit axle will do, but you have to insert it half-way into the knob wheel.

STEP 19 Create a mirror image of the previous assembly.

STEP 20 Place bushes on the axles right before inserting them into the motors.

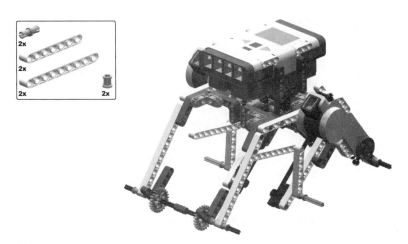

STEP 21 Rotate the model and add the beams to the middle axles as shown.

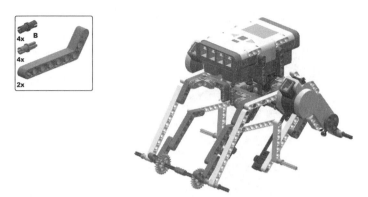

STEP 22 Join the linkage to the L-beams as shown.

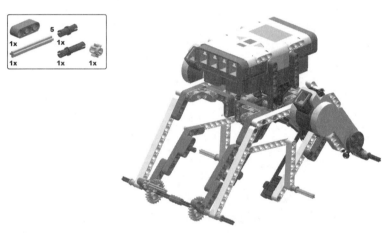

STEP 23 Mesh an 8-tooth gear with the 24-tooth gear and an axle. For now it hangs loosely. Add the remaining parts as shown.

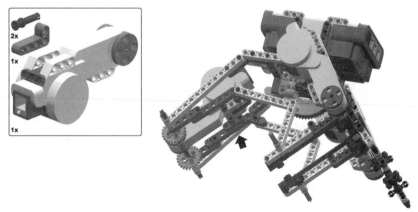

STEP 24 Place the L-beam in the center slot of the motor and secure with two long-pins with bush. The motor attaches to the axle from the previous step. Secure by connecting the pin to the L-beam.

STEP 25 Attach the wheels as shown. The rear wheels are from the RIS kit. If you don't own this kit, substitute these tires with the large grey gear and the large black double-bevel gear.

This robot uses all medium cables. Attach cables from ports B and C to the left and right motors. Attach a cable from the rear motor to port A. I positioned the cables freely above the NXT.

As you can see, this LEGO robot is a real monstrosity. You might wonder why the middle tire gears shoot past the tire, placing the gear closer to the ground. This was used because it needs to reverse the direction of the tire rotation; otherwise it will turn counter to the front wheels.

Programming Rocker-Bogie

Rocker-Bogie has the simplest code of all the robots in the book. It just starts the wheels turning and waits for you to push the orange button to end the program.

```
1.  import lejos.nxt.*;
2.
3.  public class RockerBogie {
4.    static final int SPEED = 800;
5.    public static void main(String [] args) throws
          Exception {
6.      // Front = 8:1
7.      // Rear = 3:1
8.      // Front wheels = 56, rear wheels = 49.6
9.      int ASPEED = (int)(((SPEED / 8) * 3) * (56F / 49.6));
10.     Motor.A.setSpeed(ASPEED);
```

```
11.        Motor.B.setSpeed(SPEED);
12.        Motor.C.setSpeed(SPEED);
13.        Motor.C.backward();
14.        Motor.B.forward();
15.        Motor.A.forward();
16.        while(!Button.ENTER.isPressed()) {}
17.        Motor.A.flt();
18.        Motor.B.flt();
19.        Motor.C.flt();
20.    }
21. }
```

The main interest in this code is calculating the proper speed for the rear tire rotation, since it is using a different tire size and gear ratio. The value for ASPEED is calculated by adjusting for the gear ratio and tire size, as shown in the first line of main(). This should keep the tires from skidding too much while it travels.

TIP: *When transporting Rocker-Bogie, pick it up by the NXT brick.*

Operating Rocker-Bogie

Your first thought with the Rocker-Bogie vehicle might be that it is flimsy. It's not. The flimsy feeling arises because when you pick it up, everything flexes as though nothing is solid. The suspension is designed to flex to keep all the wheels on the terrain at the same time. Let's see what it can do.

Our first test will use the six-wheel drive to overcome some steps. I used a large World Atlas. As you can see, the suspension is capable of climbing a vertical obstacle (see Figure 20-7). It helps if the wheels contact the obstacle at the same time.

Figure 20-7 Rocker-Bogie climbing up a step

Now let's remove the axle joiners on the rear axle and see how well the suspension operates on uneven terrain. As the robot drives over different-sized objects, it keeps all the wheels on the ground. This results in the robot being able to overcome all kinds of terrain. Rocker-Bogie's ability to adapt to extreme variations in terrain is impressive (see Figure 20-8).

Figure 20-8 Rocker-Bogie suspension on uneven terrain

Speech

Topics in this Chapter

- Speech Synthesis
- Speech Recognition
- Number 0.5 – A Robot That Communicates

Chapter **21**

S peech is a defining aspect of being human. Giving our robots the power of speech brings them one step closer to being just like us. This chapter will explore ways to give your robot a voice and make it understand your voice. By the end of the chapter you will have a robot that speaks and responds to your commands.

Speech Synthesis

Speech synthesis has been around for decades and can be programmed in less than 32 kilobytes of memory. However, to sound reasonably human, speech synthesis software requires a lot of memory. For this reason, the heavy lifting for speech synthesis will take place on a PC, which will then send the speech to the NXT.

There are plenty of good commercial speech synthesizers but for our purposes we need something that is platform independent, free, and good quality. FreeTTS (Free Text To Speech) fits these criteria. It is written entirely in Java and provides decent voice synthesis. FreeTTS comes with several different voices. One voice, called *Allen*, sounds almost human, but it is limited to speaking time information. The only available voice at present for general speech is called *Kevin*, and it sounds more like the Borg from *Star Trek*. The FreeTTS developers say FreeTTS is capable of very realistic voices and have promised better voices in the future, but nothing has materialized so far. Let's give it a try!

Installing FreeTTS

 NOTE: *These instructions are for version 1.2.x. If you download a later version, follow the instructions from FreeTTS.*

1. Download the software from freetts.sourceforge.net. You want the distribution with the word BIN in the file name. e.g. freetts-1.2.1-bin.zip
2. Unzip the file into a directory.

3. Set your CLASSPATH variable to include lib\freetts.jar. You must include the full path.

e.g. c:\program files\freetts\lib\freetts.jar

In Windows you can find the classpath by right-clicking My Computer > Properties > Advanced > Environment Variables.

4. Set up the Java Speech API. This is done by changing to the lib directory and typing: jsapi.exe (Windows) or sh ./jsapi.sh (Linux and Macintosh). Agree to the license and it will install.

5. Copy the speech.properties file to your user directory:

Windows: c:\Documents and Settings\User

Linux: /home/userid or /users/userid

MacOSX: /users/userid

Once FreeTTS is installed you should test the samples that are included in the BIN directory. To run them from the command line, type:

java –jar HelloWorld.jar

Before moving on, you need to download a program that will convert a file to the LEGO audio format (.RSO). The program is called WAVRSOCVT, a command line tool developed by John Hansen, available from:

http://bricxcc.sourceforge.net/utilities.html

Unzip it into a directory and remember where you put it. You will need to type this path into your code later on.

Programming Speech

In this project, we will feed different strings to the program and it will play the speech from the NXT brick. We will use iCommand in concert with the LEGO firmware. The program performs the following steps:

1. Accepts a string and uses FreeTTS to output a WAV file.

2. Executes WAVRSOCVT to convert the WAV file to an RSO file.

3. Uploads the RSO file to the NXT, then deletes the RSO file from the PC.

4. Plays the RSO through the NXT speakers, then deletes the RSO file from the NXT brick.

```
1. import icommand.nxt.comm.NXTCommand;
2. import icommand.platform.nxt.FileSystem;
3. import icommand.platform.nxt.Speaker;
4.
5. import java.io.*;
6.
7. public class SpeechSynthesis {
8.     static File freettsdir = new File("C:/Java/freetts-1.2.1");
9.     static File converttool = new File("C:/Java/wavrsocvt.
          exe");
10.
11.     static String WAVNAME = "temp.wav";
```

```
12.      static String RSONAME = "temp.rso";
13.
14.      public static void say(String sentence) {
15.
16.          String execString = "java -jar lib/freetts.jar -
                dumpAudio " + WAVNAME + " -text ";
17.          try {
18.              Process p = Runtime.getRuntime().exec(execString
                    + sentence,null,freettsdir);
19.              p.waitFor();
20.              p = Runtime.getRuntime().exec(converttool
                    + " " + WAVNAME + " -O=" + freettsdir,null,
                    freettsdir);
21.              p.waitFor();
22.          } catch (Exception e) {
23.              System.out.println("Exceptioon: " + e);
24.          }
25.          File rsoFile = new File(freettsdir + "/" + RSONAME);
26.          byte status = FileSystem.upload(rsoFile);
27.          if(status == 0)
28.              System.out.println("File successfully uploaded");
29.          else {
30.              System.out.println("Error uploading " + rsoFile.
                    getName() + ": " + Integer.toHexString(status));
31.              return;
32.          }
33.          rsoFile.delete();
34.          Speaker.playSoundFile(rsoFile.getName());
35.          NXTCommand.setVerify(true);
36.          do {} while(FileSystem.delete(rsoFile.getName())!=0);
37.      }
38.
39.      public static void main(String[] args) {
40.          NXTCommand.open();
41.          SpeechSynthesis.say("PLEASE PUT DOWN YOUR WEAPON");
42.          SpeechSynthesis.say("YOU HAVE 20 SECONDS TO
                COMPLY");
43.          SpeechSynthesis.say("YOU NOW HAVE 15 SECONDS TO
                COMPLY");
44.          NXTCommand.close();
45.      }
46. }
```

Before running the code, make sure the NXT is set to full volume. The iCommand code takes a moment to connect and then generates and uploads the speech file. You will hear it say three famous ED-209 lines from the film Robocop. It's a little quiet but audible. The program can take several seconds to generate each new line of speech, depending on the speed of your computer.

WARNING: *If you decide to use this class in your robots, keep the individual sentences short because the sound files occupy a significant amount of memory. Anything over 10 words should be broken up into shorter sentences.*

Speech Recognition

There are many speech recognition packages available, both commercial and open source. The big commercial packages are Microsoft Speech (through Microsoft Office), IBM ViaVoice, and Dragon Naturally Speaking. These packages offer excellent recognition after some training, but they are not available for all platforms.

The two major open source speech recognition packages are Sphinx and Julius. In this section we will use Sphinx, since Julius is available only for Linux and Windows and has no Java interface (it relies on command-line binaries). There are several branches of Sphinx, but we are interested in Sphinx-4, the 100% Java implementation of Sphinx.

Sphinx was developed at Carnegie Mellon University, with help from Sun Microsystems, Mitsubishi, and Hewlett-Packard. It is a limited domain speech recognition package, which means it can only identify words that appear in a small list you provide for it. This is different from commercial packages, which can recognize words from a large vocabulary. The commercial packages are good for general dictation, while limited domain speech recognition like Sphinx is good for applications like phone menus and controlling robots with commands.

TIP: *To give you an idea of how well the Sphinx voice recognition software works, you can try it out by dialing a room booking service that uses it. Dial 1-877-268-7526 (toll free) or +1 412 268 1084. When it answers, say "Book a room" or something similar, then it will proceed to ask you questions.*

Installing Sphinx

1. Download the latest Sphinx-4 release from: http://sourceforge. net/projects/cmusphinx

   Click the Download tab, then Sphinx-4. Download the file with bin in the filename, not src.

2. Extract the contents into a folder.

   e.g. c:\speech\

3. Add sphinx4\lib\sphinx4.jar to your class path.

   e.g. c:\speech\sphinx4\lib\sphinx4.jar

4. Add the class path to the Wall Street Journal voice files in the lib directory:

e.g. c:\speech\sphinx4\lib\WSJ_8gau_13dCep_16k_40mel_130Hz_6800Hz.jar

5. To check your classpath, type echo %classpath%. You should see something like the following:

    .;c:\speech\sphinx4\lib\sphinx4.jar;c:\speech\sphinx4\lib\WSJ_8gau_13dCep_16k_40mel_130Hz_6800Hz.jar

6. You also need to run the setup for the JSAPI (Java Speech API). Change to the sphinx-4/lib folder.

7. Windows users run a program called jsapi.exe. Linux users should type sh ./jsapi.sh. Once you agree to the license it will install JSAPI.

8. If you are using Eclipse you will need to add the two jar files above plus jsapi.jar in the same location. Select Project > Properties > Java Build Path to change these settings.

Testing Sphinx

Now that you have Sphinx installed, let's give it a try. The first thing to do is ensure that your microphone is connected and working. Test the microphone levels in your operating system configuration.

Windows

1. Select Start > Control Panel > Sounds and Audio Devices

2. A window will pop up with Sounds and Audio Devices Properties. Click the Voice tab (see Figure 21-1).

Figure 21-1 Setting the audio devices properties

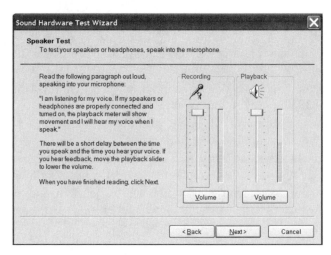

Figure 21-2 Setting the recording levels

3. Click **Test Hardware** and follow the instructions. It will show sound levels for the microphone and allow you to adjust the recording volume (see Figure 21-2).

Linux:

1. Different versions of Linux use different applications to control sound. Under Kubuntu, select K Menu > Multimedia > KMix.

2. Click the input tab and look at the indicator for Mic. If it is green, the microphone is recognized and working properly. Make sure the recording volume is not too low.

Mac OS X:

1. Click on the Apple Menu and select System Preferences.

2. The System Preferences window will appear. In the System Preferences window, click on the Sound icon.

3. The Sound preferences will appear. Click the Input tab (see Figure 21-3).

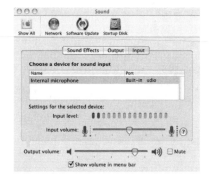

Figure 21-3 Adjusting the microphone levels

4. You should see a list of available ways to input sound such as Internal microphone or External microphone. Choose the appropriate method you will be using and make sure the volume is high enough.

Now that your microphone is set up properly, let's try a test program in Sphinx.

1. Change to the directory where Sphinx is installed.
2. We will try the HelloWorld test. The Java VM needs additional memory for Sphinx, so add the following memory parameter when running the sample:

    java -mx312m -jar bin/HelloWorld.jar

3. You may now say a greeting followed by a name in the list. Sphinx will repeat the words you say on screen.

WARNING: *The results may depend on the microphone and volume level used.*

Building Number 0.5

Movie robots have always inspired robot builders. When people dream of building a robot companion, Number 5 seems to come up quite often; he can do everything a robot is supposed to do. He's highly mobile. He can manipulate objects with great dexterity. He can talk. He exhibits intelligence. He's helpful. Other robots in movies have exhibited these traits, but they were usually men in robot costumes. Number 5 is purely mechanical, and more like a real robot. Perhaps that explains his popularity among robotics enthusiasts.

Number 5 is a good robot for demonstrating speech capabilities because he is so vocal in films. He's almost ideal for the NXT kit because the ultrasonic sensor looks exactly like his mantis-like head. Number 5 would have been more fully realized if LEGO had included tank treads in the kit, but the dual wheels work almost as well on a practical level. We'll call this robot Number 0.5 because he's about 1/10th scale.

Goals:

- Scale representation of Number 5.
- Split/Independent suspension with rear caster wheel

STEP 1 Insert the yellow axle-pin into the dark grey axle hole. Attach the ligh-grey connector to the yellow axle-pin so it rotates freely. Insert the axle in the light-grey axle hole.

STEP 2 Add parts as shown.

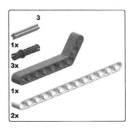

STEP 3 Add parts as shown.

STEP 4 Add parts as shown.

STEP 5 Add parts as shown.

STEP 6 Stagger the long pin with a short pin on each side of the L-beam.

STEP 7 Add parts as shown.

STEP 8 Add parts as shown.

STEP 9 Add parts as shown.

STEP 10 Add parts as shown.

STEP 11 Add parts as shown.

STEP 12 Add parts as shown.

STEP 13 Add parts as shown.

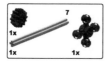

STEP 14 Add parts as shown.

STEP 15 Add parts as shown.

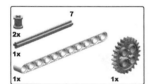

STEP 16 Add parts as shown.

STEP 17 Add parts as shown.

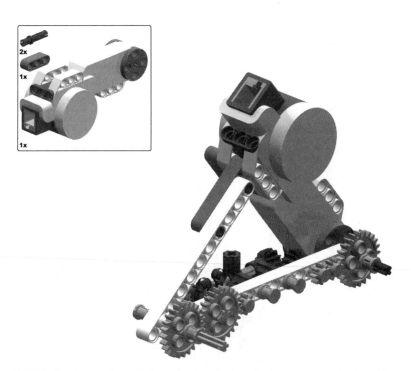

STEP 18 Insert the axle into the motor. Attach the motor on the far side using double pins, leaving the top hole open (see step 20 for reference). Attach a 3-unit beam.

STEP 19 Add parts as shown.

STEP 20 Repeat a mirror image for the other side.

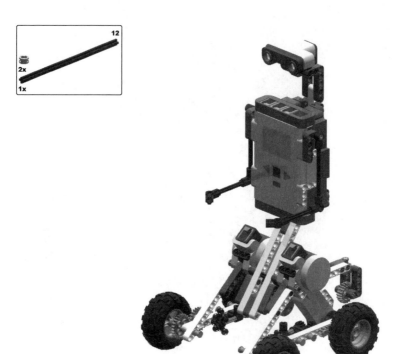

STEP 21 Attach the drive mechanism by inserting an axle through both sides, as shown, securing with half bushes.

Use 3 medium sized cables. Thread the cables between the slot in the middle of the robot. The cable from the ultrasonic sensor leads to port 1, and the two motors connect to ports B and C.

Programming Number 0.5

Sphinx requires a separate grammar file (.gram) containing a list of commands. The following is an example of a grammar file:

```
 1.  #JSGF V1.0;
 2.
 3.  /**
 4.   * JSGF Grammar for Robot control
 5.   */
 6.  grammar commands;
 7.  public <orders> = ( Stop | <turn> | <move> | Quit |
        Open);
 8.  <move> = (Go <throttle>);
 9.  <throttle> = (Forward | Backward | Back);
10.  <turn> = turn <direction>;
11.  <direction> = (left | right);
```

The grammar file has a specific syntax, much like a programming language. Some words always follow another word, such as go and forward. The grammar file lets you specify the order in which these words are received, as shown above for the move command.

Sphinx also requires a long and complex configuration file. You can download the configuration file (along with the code) from the publishers website listed at the front of this book, or you can use one of the sample configuration files (such as Hello World). If you do this, you must change a line in the file under the Grammar configuration as follows:

```
<property name="grammarName" value="commands"/>
```

TIP: *If you decide to make your own commands, longer multi-word commands like "turn-left" and "go forward" are actually easier for Sphinx to recognize than single word commands like "left" or "stop".*

Before we begin coding Number 0.5 we need to decide what commands he will obey. Table 21-1 lists these commands and the resulting action.

Voice Command	Action
"Stop"	Stops
"Go"	Drives forward
"Backward"	Drives backward
"Left"	Turns left
"Right"	Turns right
"Faster"	Speed increases by 10
"Slower"	Speed decreases by 10

Table 21-1 Commands for Number 0.5

The grammar file for Number 0.5 is simple:

```
1.  #JSGF V1.0;
2.  /**
3.   * JSGF Grammar for Number 0.5
4.   */
5.  grammar commands;
6.  public <orders> = ( Stop | Go | Backward | Left | Right
        | Faster| Slower| Quit);
7.  Save the above file as commands.gram in the same
        directory as the Java classes. The code below uses a
          mixture of Sphinx classes and iCommand to control
            Number 0.5:
8.  import edu.cmu.sphinx.frontend.util.Microphone;
9.  import edu.cmu.sphinx.recognizer.Recognizer;
10. import edu.cmu.sphinx.result.Result;
11. import edu.cmu.sphinx.util.props.ConfigurationManager;
12. import edu.cmu.sphinx.util.props.PropertyException;
13.
14. import java.io.*;
15. import java.net.URL;
16.
17. import icommand.nxt.*;
18.
19. public class VoiceControl {
20.
21.   public static void main(String[] args) {
22.
23.     Pilot robot = new Pilot(5.6F, 16F, Motor.B, Motor.C);
```

```
24.
25.      try {
26.        URL url;
27.        if (args.length > 0) {
28.          url = new File(args[0]).toURI().toURL();
29.        } else {
30.          url = VoiceControl.class.getResource("voice.con
               fig.xml");
31.        }
32.
33.        System.out.println("Loading...");
34.
35.        ConfigurationManager cm = new Configuration
            Manager(url);
36.
37.        Recognizer recognizer = (Recognizer)
             cm.lookup("recognizer");
38.        Microphone microphone = (Microphone)
             cm.lookup("microphone");
39.
40.        recognizer.allocate();
41.
42.        icommand.nxt.comm.NXTCommand.open();
43.
44.        if (microphone.startRecording()) {
45.
46.          System.out.println("Commands: ");
47.          System.out.println("Go, Stop - controls robot
                 automatic AI program.");
48.          System.out.println("Left, Right, Forward,
               Backward, Back, Stop - direct robot control");
49.          System.out.println("Faster, Slower -
               control speed");
50.
51.          while (true) {
52.            System.out.println("Start speaking. Say QUIT to
                 quit.\n");
53.
54.            Result result = recognizer.recognize();
55.
56.            if (result != null) {
57.              String resultText = result.
                   getBestFinalResultNoFiller();
58.              System.out.println("You said: " + resultText
                 + "\n");
59.              if(resultText.equals("quit")) {
60.                icommand.nxt.comm.NXTCommand.close();
61.                System.exit(0);
62.              }  else if(resultText.equals("faster"))
63.                robot.setSpeed(robot.getSpeed() + 10);
```

```
64.            else if(resultText.equals("slower"))
65.               robot.setSpeed(robot.getSpeed() - 10);
66.            else if(resultText.equals("go"))
67.               robot.forward();
68.            else if(resultText.equals("backward"))
69.               robot.backward();
70.            else if(resultText.equals("left"))
71.               robot.left();
72.            else if(resultText.equals("right"))
73.               robot.right();
74.            else if(resultText.equals("stop"))
75.               robot.stop();
76.         } else System.out.println("I can't hear what
            you said.\n");
77.      }
78.    } else {
79.        System.out.println("Cannot start microphone.");
80.        recognizer.deallocate();
81.        System.exit(1);
82.          }
83.  } catch (Exception e) {
84.    System.err.println("Problem loading VoiceControl: "
        + e);
85.    e.printStackTrace();
86.  }
87.  }
88. }
```

Results

If all goes well you should find yourself in control of Number 0.5.
Try to speak slowly and clearly into the microphone. This is sometimes
difficult if your robot is about to hit a wall.

TRY IT: Add the following commands to the code:
- *When you increase or decrease speed, it voices the correct
  speed using the FreeTTS speech code.*
- *Use a navigator class (and optional compass) so that you can
  tell the robot to come home to the origin location (0, 0)*
- *Make the robot semiautonomous, so that it wanders around
  by itself and avoids objects using the ultrasonic sensor, but
  at the same time can stop or go at your command.*

NXT Accessories

This section lists the parts and prices for the most popular NXT accessories. Most of these are not available at retail stores, so your best bet is to order direct from the company website. If you want to obtain parts that are no longer available, BrickLink or LUGNet are great places (see Appendix C). You can also try eBay.

NOTE: *All prices below are in US dollars. Prices may change.*

Appendix A

A1. LEGO

- www.shop.lego.com

MINDSTORMS NXT Kit (8527) $249.99
NXT Intelligent Brick (9841) $134.99
Rechargeable Battery (9798) $49.99
Transformer (9833) $22.99
Rechargeable Battery Set (K9833) $72.98
Converter Cables (8528) $9.99
Connector Cables (8529) $9.99
Interactive Servo Motor (9842) $17.99
Bluetooth Dongle (9847) $37.99
Light Sensor (9844) $16.99
Sound Sensor (9845) $24.99
Touch Sensor (9843) $16.99
Ultrasonic Sensor (9846) $31.99

LEGO Education

- www.legoeducation.com

LEGO MINDSTORMS Education Base Set (W979797) $250.00
LEGO Education Resource Set (9648) $59.00

Peeron

- www.peeron.com

9V Battery Box (2847) $14.50

Used to power extra motors for third party motor multiplexers.

A.2 HiTechnic

- www.hitechnic.com

Compass Sensor (NMC1034) $46.99

Color Sensor (NCO1038) $46.99

Tilt Sensor (NAC1040) $46.99

Extended Connector Cable Set (NWS1000) $6.49

Contains four cables of different lengths: 12, 16, 70, and 90 cm.

Prototype Board (NPB 1015)

A.3 Mindsensors.com

- mindsensors.com
  info@mindsensors.com
  Ph: +1-804-200-3376
  Fax: +1-425-984-7844

RCX to NXT Communication Adapter (NRLink-Nx) $32.00

Acceleration Sensor for NXT - (ACCL-Nx-2g2x) $39.95

Acceleration Sensor for NXT - (ACCL-Nx-3g3x) $42.95

Acceleration Sensor for NXT - (ACCL-Nx-5g2x) $39.95

Motor Multiplexer for NXT (MTRMX-Nx) $42.00

Magnetic compass for NXT (CMPS-Nx) $39.95

Sony PlayStation 2 Controller interface for NXT (PSP-Nx) $35.00

Pneumatic Pressure Sensor for NXT (PPS35-Nx) $33.00

A.4 Wireless Camera

- www.x10.com

  A wireless camera is used for projects in Chapter 10. You can purchase a new wireless system from X10 or you can find them cheaper on auction sites like eBay. A complete system requires the following (see Figure A-1).

Wireless Camera (XC10A) $79.99

Receiver (VR36A) $39.99

Receives video only (no audio).

Video to USB Adapter (VA11A) $69.99

Battery Pack (ZB10A) $19.99

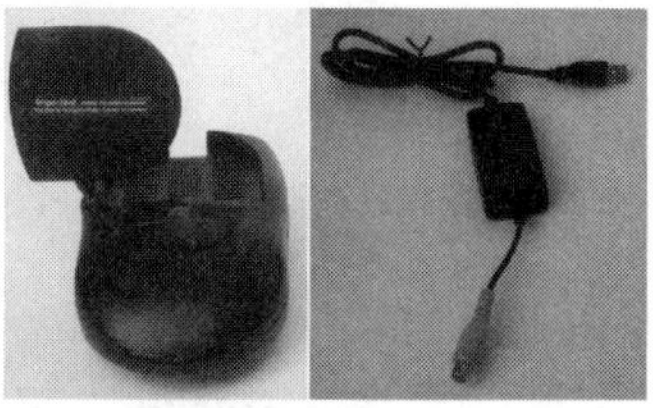

Figure A-1 X10 Receiver and Video to USB adapter

Computer Geeks

- www.geeks.com

  Computer Geeks markets a number of wireless cameras under the *Surveillance Cameras* category

A.5 Data Glove

The P5 data glove can be purchased new from eBay for as little as $10 or from a number of online retailers.

VRealities (US) $59.00

- www.vrealities.com

CyberWorld (Canada) $59.00

- www.cwonline.com

P5Glove.net - $25

- www.p5glove.net

A.6 Techno-Stuff

TechnoStuff sells a number of MINDSTORMS sensors, but there is not a lot for the NXT.

- www.techno-stuff.com

A.7 Other Robot Kits

Here are some competing robotic kits.

Vex

- www.vexlabs.com

  The Vex Starter Kit is sold in individual parts. All the parts in the kit end up costing about $300. You can also find additional parts and upgrades on their website.

fishertechnic

- www.fischertechnik.com (North America)
- www.fischertechnik.de (Europe)

Robo Mobile Set $332.99

Pneumatic Robots $189.99

Industry Robots II $219.00

Robot Math

People seem to either love math or hate it. If you love robotics, you better learn to love math. This section covers some useful mathematics that can help you program your LEGO robots.

Appendix **B**

B.1 Circumference

The distance around the outside of a wheel can be important for LEGO creations, since this can be used to calculate distance traveled (see Figure B-1). Once you know the circumference, distance is measured by the number of wheel rotations multiplied by circumference. You can calculate circumference as follows:

```
Circumference = Pi * Diameter
```

The Pi value can be found in leJOS as Math.PI.

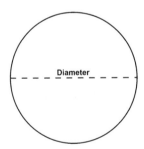

Figure B-1 Calculating circumference

B.2 Trigonometry

You are taught a lot of math in high-school, but the equations that keep coming back are the trigonometry equations. You remember: sine, cosine, and tangent. These equations go back to the time of Hipparchus, a Greek mathematician who was the first to use sine around 150 BC. Trigonometry is simply a branch of mathematics dealing with the relationships of the sides of triangles and angles.

Trigonometry is useful for calculating x,y coordinates for navigation, as well as calculating angles for arm movement in three dimensions. The Navigator classes in the leJOS API uses a lot of trigonometry to update coordinate values.

Angles can be measured in two ways. The first, most common form is degrees. A complete rotation is 360 degrees. However, engineers, scientists and mathematicians prefer using radians because the units are not arbitrarily arrived at. A complete rotation in radians is the value 2pi, or about 6.28 radians.

You can use whichever system you are more comfortable with. There are methods in the Math class for conversion between the two (see Math class in Chapter 3). In the coordinate system, zero degrees (or zero radians) always runs along the x axis, and positive rotation occurs counter clockwise (Figure B-2). Thus, when the robot is pointed north (along the Y axis) it is at 90 degrees.

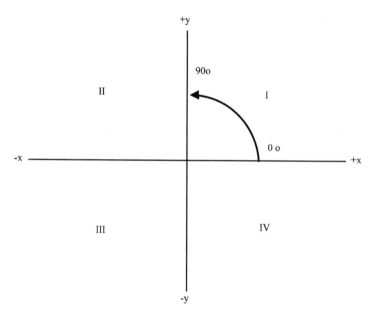

Figure B-2 Rotation

Let's examine the problem of navigation. Every time a robot moves a distance, the x and y coordinates will also change. For example, if the robot rotates 60 degrees to the left and travels twenty centimeters, both the x and y values will increase (Figure B-3). As you can see, this movement creates a triangle with a right angle. In trigonometry, when there is a right angled triangle with all angles known, and the length of one side is known, it is possible to calculate the lengths of all sides.

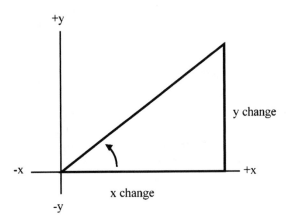

Figure B-3 Right angle trigonometry

In trigonometry, the side of the triangle opposite the right angle is called the hypotenuse, the side opposite the angle in the calculation is the opposite, and the remaining side is the adjacent (Figure B-4). In order to calculate the new location we will need to calculate x (the adjacent) and y (the opposite). To do this, we will need to use calculus.

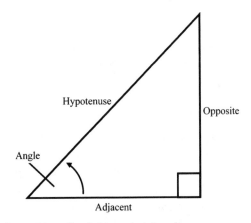

Figure B-4 The three sides of a right angle triangle

In high school calculus you probably learned tangent, cosine, and sine (tan, cos, and sin). These functions are used to solve for the lengths of sides on a triangle. Many people use the mnemonic SOH, CAH, TOA (say it like a tribal chant) to remember the following equations:

```
Sin(angle) = Opposite/Hypotenuse
Cos(angle) = Adjacent/Hypotenuse
Tan(angle) = Opposite/Adjacent
```

We only need to know the opposite and adjacent, so only the first two equations are useful to us. Let's replace the technical terms with variables and rearrange the equations to make things simpler. The distance traveled by the robot, the hypotenuse, will be replaced by distance:

```
x = cos(angle) * distance
y = sin(angle) * distance
```

We can now use these equations to figure out the x and y coordinates after a robot has moved a distance across the floor. Let's imagine the robot has started at 0,0 and rotates positive 70 degrees (counter clockwise), then moves 25 centimeters (Figure B-5). In order to find new coordinates simply plug our values into the equations above to get:

```
x = cos(70) * 25
y = sin(70) * 25
```

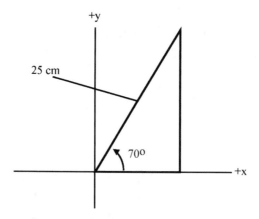

Figure B-5 Rotating 70 degrees and traveling 25 cm

If you are using a calculator, make sure it is in degrees (DEG mode) and not radians (RAD mode). So the new x and y coordinates are 8.55 and 23.49. In Java, it is very easy to calculate these using Math class:

```
double x = Math.cos(Math.toRadians(70)) * 25;
double y = Math.sin(Math.toRadians(70)) * 25;
```

NOTE: *All methods in the java.lang.Math class use radians for angles. There are two methods available to convert back and forth between degrees to radians: Math.toRadians() and Math.toDegrees().*

B.3 Inverse Trigonometry Functions

In the preceding section we knew distances and angles, and used those to calculate x and y coordinates. However, sometimes we know the x and y coordinates and must calculate the angle. This is the problem faced when programming robot arms (see Chapter 8).

In this case we use inverse functions. Take a look at figure B-4 again. If we know the length of each side of a triangle (x, y coordinates) then we can calculate the angles by rearrange the equations slightly:

```
angle = asin(Opposite/Hypotenuse)
angle = acos(Adjacent/Hypotenuse)
angle = atan(Opposite/Adjacent)
```

Asin, acos and atan are merely words for inverse sine, inverse cosine, and inverse tangent. In Figure B-5, pretend we know the x, y coordinates but don't know the angle indicated. We can use atan by entering the x and y values:

```
Angle = atan(y/x)
Angle = atan(23.49/8.55)
Angle = 69.99 (rounded to 70)
```

There is one problem with atan—it can't calculate angles greater than 90 degrees. However, using Math.atan2(y, x) you can calculate this accurately as atan2() can produce any angle between 0 and 360 degrees (or more accurately, between 0 and 2pi).

B.4 Law of Cosines

Sometimes you don't have a right angled triangle. In this case, as long as you know the lengths of all three sides of a triangle, you can still figure out the angles. In the arm project in chapter 8 we know the lengths of two parts of the arm (two sides) and we can calculate the distance between the ends of each arm. This means we can calculate the inner angle (see Figure B-6).

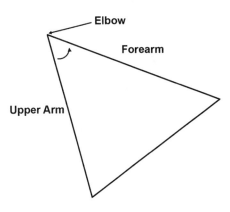

Figure B-6 Robot arm triangle

The law of cosines applies to any triangle in which the lengths of all three sides are known (see Figure B-7). The law is as follows:

$$c^2 = a^2 + b^2 - 2abcosA$$

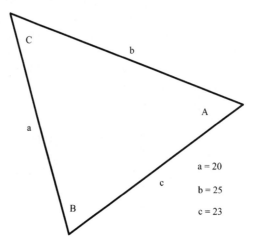

Figure B-7 Law of cosines

In this form the equation isn't useful if we want to calculate angle A, so we rearrange as follows:

$$A = acos\left(\frac{a^2 + b^2 - c^2}{2ab}\right)$$

Now we merely plug in the values for a, b and c and we can determine the angle:

$$A = acos\left(\frac{20^2 + 25^2 - 23^2}{2 \times 20 \times 25}\right)$$

$$A = acos\ 0.496$$
$$A = 60.26\ degrees$$

Internet Resources

Usually you can find what you are looking for easily with a simple Google search. This section will only concentrate on premier sites that offer a specific LEGO NXT service.

Appendix C

leJOS

- www.lejos.org

  This website is the main headquarters for leJOS. You can download new files, chat about leJOS in the forums, read news, browse the API documentation, read tutorials and more.

Maximum LEGO NXT

- www.variantpress.com

  The official website for this book. Download code listings for the entire book, find errata and updates.

NXTasy

- www.nxtasy.org

  NXTasy is one of the foremost sites to read all the latest NXT news. It also has a projects section and some good forums.

LEGO MINDSTORMS

- mindstorms.lego.com

  Here you can find projects that other users have created, new sensors and parts, and a list of supported Bluetooth dongles.

BrickLink

- www.bricklink.com

  BrickLink is the unofficial LEGO marketplace; sort of like eBay for LEGO. Here you can purchase hard to find parts and kits from all around the world.

LUGnet

- www.lugnet.org

  LUGnet stands for LEGO User Group Network. Here you can find a variety of newsgroups to chat with other users about your favorite toy.

YouTube

- www.youtube.com

What's YouTube doing in a list for NXT? As it turns out, NXT owners like to upload videos of their creations to YouTube. You can find most of them from NXTasy.org in the Repository section under videos or do a YouTube search for NXT.

BrickCC

- www.bricxcc.sourceforge.net/utilities.html

BrickCC is another development environment for NXT but the site also includes some useful tools by John Hanson. These tools allow you to explore the NXT brick or convert sound files to NXT format.

LDraw

- www.ldraw.org

This website is the home base for LDraw, a brilliant program that allows users to model their creations in a CAD program. If you design a nice model and want to save the plan, this utility is for you.

Index